PENCIL
NOODL
CUP
BEER
PENCIL

KITTY
CHICKEN
LITTER

BRIEF HISTORIES OF EVERYDAY OBJECTS

BRIEF HISTORIES OF EVERYDAY OBJECTS

Written and Drawn by
Andy Warner

Picador

New York

For information, address Picador, 175 Fifth Avenue, New York, N.Y. 10010.

picadorusa.com
picadorbookroom.tumblr.com • twitter.com/picadorusa • facebook.com/picadorusa

Picador® is a U.S. registered trademark and is used by Macmillan Publishing Group, LLC, under license from Pan Books Limited.

For book club information, please visit facebook.com/picadorbookclub
or e-mail marketing@picadorusa.com.

Illustrated by Andy Warner

The Library of Congress Cataloging-in-Publication Data is available upon request.

ISBN 978-1-250-07865-0 (paper over board)
ISBN 978-1-250-07866-7 (e-book)

Our books may be purchased in bulk for promotional, educational, or business use. Please contact your local bookseller or the Macmillan Corporate and Premium Sales Department at 1-800-221-7945, extension 5442, or by e-mail at MacmillanSpecialMarkets@macmillan.com.

First Edition: October 2016

10 9 8 7 6 5 4 3 2 1

For Mr. Gonick and Uncle John

CONTENTS

THE COFFEE SHOP
Tea ~ pg. 90
Artificial Sweeteners ~ pg. 94
Cinnamon ~ pg. 98
Coffee Beans ~ pg. 102
THE OFFICE
Paper Clips ~ pg. 108
Ballpoint Pens ~ pg. 112
Paper ~ pg. 116
Pencils ~ pg. 120
Post-it Notes ~ pg. 124
Stamps ~ pg. 128
GROCERIES
THE GROCERY STORE
SALE! 50%
Paper Bags ~ pg. 134
Instant Ramen ~ pg. 138
Canned Fruit ~ pg. 142
Ice Cream Cones ~ pg. 146
Potato Chips ~ pg. 150
THE BAR
Beer Cans ~ pg. 156
Toothpicks ~ pg. 160
Ice ~ pg. 164
Billiard Balls ~ pg. 168
THE GREAT OUTDOORS
Traffic Lights ~ pg. 174
Roller Skates ~ pg. 178
Streetlights ~ pg. 182
Barbed Wire ~ pg. 186
Kites ~ pg. 190
Bicycles ~ pg. 194
BIBLIOGRAPHY ~ pg. 199
ACKNOWLEDGMENTS ~ pg. 203
INDEX ~ pg. 205

INTRODUCTION
Brief Histories of Everyday Objects had its start in a shower I took in late February of 2013.
I was trying to come up with an idea for a comic.
Ok, next time I have the coffee before doing this....
Maybe showerheads? Where did they come from? There could be a story in showerheads!
There was no story in showerheads.
drip!
drip!
They're actually pretty boring.
My toothbrush, however, turned out to hide a history of riots, trading routes and London prisons.
Ith amathingh!!!
Shooga shooga
I kept digging. All around me, from bags to ballpoint pens, I found stories.
BEER
NET WT
POTATO CHIPS
Mayflower
TASTY
FRUIT COCKTAIL
I realized that I had my comic idea.

The history in this book is as true as I could get it, but I made up the dialogue wholesale because it's funnier that way.
This probably needs more vacuum cleaner jokes.
CHEW CHEW
When I quoted someone directly, I made sure to make a note of it.
Whenever I couldn't find any visual reference for what people or places actually looked like, I winged it. Hopefully, nobody gets too mad....
cough
Then, for the past year, I set up studio in a garden shed and drew everything.
A year of my life On toothbrushes. Man.
It's a pretty decent shed, though. I hope you enjoy the comics!
BRIEFER HISTORIES
I came across a lot of great things in my research that I didn't use in the stories.
These "Briefer Histories" sections are where I stuck that stuff.
In case you want more trivia about paper clips!
Because everybody wants more paper clip trivia.
Oh god, I hope so.

THE BATHROOM
Toothbrushes pg. 2
Shooga shooga
Shampoo pg. 6
Razors pg. 10
Toilets pg. 14

Bathtubs
pg. 18
Kitty Litter
pg. 22

TOOTHBRUSHES
Dental hygiene practices in 18th-century England were not particularly effective.
Most people just rubbed rags coated with soot and salt on their teeth.
Ack!
Gross!
Why are we doing this?
Then, in 1770, a man named William Addis was sent to a London jail for inciting a riot.
We were just blowing off a little steam!
OW!!!

Addis suddenly found himself with a whole lot of time to kill.
Sigh...
Guess I'll brush my teeth with the old soot-rag again.
Addis whiled away the hours by turning his mind to invention. He drilled holes in a bone left over from his dinner, stuck in animal hair bristles...
fiddle fiddle fiddle
Everybody needs a hobby.
...and managed to invent a toothbrush!

But Addis might not have come up with the idea entirely on his own.
SHOOGA SHOOGA SHOOGA
SHOOGA SHOOGA
SHOOGA SHOOGA
Toothbrushes had been used in China since at least 700 AD.
They'd been imported to Europe since at least the 16th century.
MYSTERIES FROM THE EAST
Addis might very well have seen them around.
It's impossible to know if Addis swiped the idea from ancient China, but it almost doesn't matter.
I'm gonna incite a riot for dental hygiene!
fiddle fiddle fiddle fiddle fiddle fiddle
Once he got out of the slammer, Addis founded the world's first mass-produced toothbrush company...

BRIEFER HISTORIES

Most Americans only started brushing their teeth when soldiers brought the habit back with them after World War II.

An Oakland photographer named Maryly Snow has over 1,000 toothbrushes in a collection she's been amassing since 1981.

Boar and badger hair were used for bristles until nylon took over in 1938. Napoleon was fancy. He used horse hair for *his* toothbrush.

Ingredients in ancient toothpaste included ox hooves, eggshells, oyster shells, and charcoal. Minty fresh!

SHAMPOO
Sarah Breedlove was born in 1867 to parents who'd escaped slavery only two years previously. Life was hard.
Oh, you mean it might be difficult to get ahead after being bought, sold and forced into unpaid labor since you were born?
I'm shocked!
Sarah's parents died before she was 8. She was married by 14.
Really just leaping into the narrative here, huh?
At 18, she gave birth to a daughter. Then her husband died, too, leaving Sarah a widowed, orphaned single mother at the age of 20.
Well, at least things can't really get any worse.
WAAAH!

Then Sarah's hair started falling out.
Sigh...
In 1904, Sarah came across a product marketed by a woman named Annie Turnbo Malone. It healed her hair!
Imagine that!
PORO
Something that's not a horrifying tragedy!
Sarah was impressed. She got a job as a sales agent for Turnbo's products.
Sarah's love life turned around, too. She married a salesman named Charles Joseph Walker and took his name.
She was now "Madam C. J. Walker."
The newly minted Madam Walker was visited by an African man in a dream.
It's ok. I'm here to tell you hair-care secrets....
Inspired, Madam Walker decided to sell her own hair formula. Malone never forgave her.
Whatever.
I'm really gonna give this whole "not horrifying tragedy" thing a shot.
MADAM C.J. WALKER'S
TRADE MARK REGISTERED
VEGETABLE SHAMPOO
A SPLENDID FOOD FOR DANDRUFF AND SORE SPOTS FOR SCALP AND SKIN
PRICE 50 CENTS
At the time, many African American women were using goose fat to style their hair.
"Better than goose fat."
I should write that down.
Madam Walker's product proved popular.

Walker recruited a dedicated sales force of other young African American women and paid them well.
Her well-coiffed saleswomen wore matching uniforms and sold Madam Walker's products door to door.
"Better than goose fat!"
She trained an army of tens of thousands of saleswomen.
Our own hair is half the sale!
Madam Walker's marriage fell apart. She kept the name.
I've seen worse. I've really, really seen worse.
Business boomed.
She put her wealth to good use. When a cinema forced her to pay a higher price because of her race, she sued...
TICKETS
Ooh, they are going to *regret* this.
WALKER THEATRE
...and then built an enormous entertainment complex to cater to African Americans, complete with a beauty salon, drugstore, ballroom, coffee shop and 1,500-seat theater.
OK, she was right about the regret thing....
WALKER
Beauty Shop
COFFEE POT
SODA
DRUGS
CIGARS
DRUGS
SODA

Madam Walker used her money and influence to fight lynchings and promote black-run businesses.
N·A·A·C·P
Anti-Lynching Campaign
W. E. B. Du Bois!
She built a mansion near the Rockefellers' estate...
...only to die aged 51, from kidney failure.
Upon her death, the orphaned, widowed daughter of ex-slaves was estimated to be America's first self-made female millionaire.
Not a bad run, all things considered.
MADAM C.J. WALKER'S
VEGETABLE SHAMPOO
BRIEFER HISTORIES
Annie Malone — who Madam Walker had worked for and gotten the idea for hair care products from — was also the orphaned child of ex-slaves.
A lot of that going around in 1860s America.
Malone used the chemistry she'd picked up before dropping out of school to develop her product.
Slightly more reliable than a dream about an African dude.
Malone became a millionaire a few years after Walker, only to lose much of her money in a divorce from a bible salesman.
But not before I bought an entire city block in Chicago.
Annie Malone lived to the ripe old age of 87.
An. Entire. City. Block.
Boom.

RAZORS
King C. Gillette, the man who made the world clean-shaven, was inspired by a bottle cap.
Pop
Hm.
Gillette was the protégé of William Painter, a bottle-cap tycoon.
Painter was an Irish immigrant to the United States who'd convinced American bottle makers to adopt a universal bottle cap for their bottles.
And guess who owns the patents on the bottle caps and bottle openers?
W. PAINTER.
BOTTLE SEALING DEVICE.
No. 468,226.
Fig. 2.
Fig. 1.
Fig. 4
Fig. 3.
Painter shared the secrets of his success with Gillette.
Invent something disposable, kid.

One morning, while shaving himself with a straight razor in the mirror, Gillette had an idea.
SLIP!
Something... disposab -ow!
Ok, next time I have the coffee *before* doing this.
Gillette partnered with a machinist named Nickerson to develop a shaving solution.
OK — still some kinks to work out...
In 1901, they invented the safety razor and disposable razor blade, then founded a company named "Gillette" to market them.
I suggested "Nickerson," but nooo...
Gillette's mustache and clean-shaven chin became the face of the brand.
I Want You to Try My Razor
AN IDEAL HOLIDAY GIFT
Gillette Safety Razor
NO STROPPING NO HONING
They sold 168 razor blades their first year. The following year they sold over 123,000.
We... uh... hit sales targets.

Gillette suddenly found himself with the money to finance his even grander dreams.

In 1902, he published a book called *The Human Drift*. It was pretty heady stuff.

Gillette argued that all industry should become one giant publicly owned conglomeration...

...and that all Americans should live in a huge self-contained city in upstate New York called "Metropolis."

It would all be powered by electricity generated by Niagara Falls.

Gillette eventually set up a company to try to make his vision come to life.

He even went so far as to try to hire Theodore Roosevelt as its president, for a salary of a million dollars.

T.R. declined.

Shockingly, Gillette's utopian plans never took off and he retired, a disappointed man, to Palm Springs.

BRIEFER HISTORIES

Razors date back at least to the Bronze Age, and took all sorts of crazy forms over the centuries.

Alexander the Great had his armies clean-shaven so that their enemies couldn't grab them by their beards.

Hairless armpits for women only became fashionable in 1922, alongside sleeveless dresses. Hairless legs became popular 20 years later, when skirts got shorter.

Gillette sold the actual razors at a loss, but more than made up for it selling the disposable razor blades that they used.

TOILETS

TOILETS
The flush toilet has its origins in the innovations of a 12th-century engineer and inventor named al-Jazari.
As a young man, al-Jazari made his way to Baghdad to study in the famous "House of Wisdom."
This place seems aptly named!
Al-Jazari designed large automated mechanical contraptions, one of which was a hand-washing station.
A really, really complicated hand-washing station....
A life-sized statue of a woman filled a basin with water, which then drained by means of a hollow duck.
So practical!
psssh!
CREAK
glug 'n' glug

The mechanism he developed to make it work is now used in modern toilets to refill the water tank after a flush empties it.
Flush
But al-Jazari wasn't content to just clean hands.
Not that hygiene isn't important!
He also invented a water-powered elephant clock...
tick tick tick tick
...and created a floating band of automaton musicians.
Splash splash
His plan for a hydraulically powered city water system (the world's first!) was implemented in Damascus after his death.
Didn't get around to it!
Way less fun than a floating robot band.
But it didn't occur to anyone to stick al-Jazari's flusher on a toilet until 1596, when John Harington built a toilet at his estate in Kelston, England.
Harington was a member of Queen Elizabeth I's court (where he was nicknamed the "saucy godson"). He installed one for the monarch, too.
For her majesty's royal... uh... depositings.
cough
But the queen thought the toilet too noisy. She stuck with her chamber pot.
FLUSH!
Aiiee!

Harington decided to spread the gospel of the flush anyway. He published a pamphlet about his invention.

BRIEFER HISTORIES

Al-Jazari described 100 mechanical devices he'd built himself in *The Book of Knowledge of Ingenious Mechanical Devices* published in 1206 AD.

John Harington was banished from court another time for his translation of *Orlando Furioso*, which was deemed too racy.

The 1851 Great Exhibition held in the Crystal Palace was the first World's Fair. There were 13,000 exhibits and 6 million attendees.

The tipping point for London's sewers was the Great Stink in the summer of 1858, when sewage clogged the river Thames itself.

BATHTUBS
On December 28, 1917, readers of the New York Evening Mail were shocked to discover that they'd missed an important event!
A NEGLECTED ANNIVERSARY
Well, golly!
The article was by H. L. Mencken
That's me!
Mencken wrote that the bathtub had first been imported to America 75 years earlier, by a Cincinnati cotton dealer. It was lined with lead, weighed 1,750 pounds and was 7 feet long.
Doctors denounced it immediately.
It causes zymotic diseases!
Uh... that does sound pretty gross....

The debate over baths raged, Mencken wrote, until it was finally put to rest by President Millard Fillmore.
Pop!
Pop!
Pop!
Fillmore installed the first bathtub in the White House in 1851.
Soak it up, my fellow Americans!
The nation followed suit.
Mencken's history of the bathtub soon spread far beyond the *New York Evening Mail*, appearing in textbooks and medical histories.
esident Millard Fillmore that, even more than
val, gave the bathtub recognition and
had bought his house from the estate. Fillmore was
entertained in this house and, according to Chamberlain,
his biographer, took a bath in the tub. Experiencing no
he became an ardent advocate of the new
There was only one problem: Mencken had made the whole story up as a prank.
Wait... really?
fizzle
Sploosh.
Uh-huh.

Nine years after he wrote the article, Mencken came clean.
And I'm serious, this time.
Really!
He went on to say that "if there were any facts in it they got there accidentally."
But it was too late.
The Fillmore story had become accepted as common knowledge. In the five years following Mencken's confession, more than a dozen newspapers and magazines had reprinted the hoax as fact.
GROAN...
The story still pops up frequently today.
In truth, it was Andrew Jackson who installed the first running-water bathtub in the White House in 1833.
Nothing like a relaxing soak after a genocide...
PSSH
And porcelain bathtubs were in fact an American invention. They were advertised as having dual use as hog scalders.
KOHLER
Bath Tub & Hog Scalder
SALE!
Um... wait. Back that up for a second.

Not that H. L. Mencken would have particularly cared. In 1926, he wrote:

BRIEFER HISTORIES

Turkish baths came from Romans who got the habit from the Greeks. Turks started cleaning up their act after conquering Constantinople.

It's a myth that folks in the Middle Ages didn't bathe. But priests at the time did warn that indulging too often would lead to sin.

Archimedes discovered the physics of displacement by watching the water rise when he got in the bathtub.

Archimedes then ran naked and wet through the streets of ancient Syracuse shouting "Eureka!" That's Greek for "I have found it!"

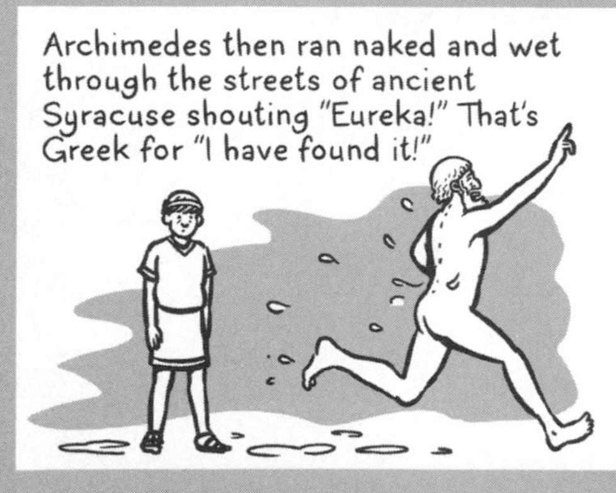

Her next-door neighbor, Edward Lowe, had his own frustrations. He'd been selling clay pellets as nesting material for hens.

The farmers weren't buying.

Draper told Lowe about her cat box problems.
...covered in ash and poop!
Have you... um.... tried clay pellets?
purrr...
CHICKEN LITTER
CHICKEN LITTER
The clay pellets were absorbent and deodorizing. Draper and her cat were immediate fans.
Well, it was this or pooping in a box of frozen sand. ...Think about it.
KITTY CHICKEN LITTER
Lowe knew he was on to something.
But he found that local stores were a little skeptical that hand-lettered 5-pound bags of clay pellets would sell.
Heck!
Give it away for free!
See if it catches on!
KITTY CHICKEN LITTER
KITTY CHICKEN LITTER
So they did.
"KITTY LITTER" FREE!
KITTY CHICKEN LITTER
KITTY CHICKEN LITTER
Curious cat owners gave Lowe's free pellets a shot...
Hey! Luxurious clay!

Kitty litter caught on, and then some.
KITTY LITTER
People even started paying money for it!
Lowe took his product on the road, crisscrossing the country cleaning cat boxes at pet shows.
CAT SHOW
It makes poop-scooping a dream!
People were intrigued.
KITTY CITY
Well, he just seemed so passionate about it.
FREE DEMO!
Pet stores took notice.
And kitty litter sales took off like a rocket.
We aim to stay No. 1 in a No. 2 business.*
*He actually said this!

Now that cats didn't track ash and poop all around the home their popularity as house pets surged.

BRIEFER HISTORIES

American cat owners fork over around $12 billion a year on kitty litter.

Kitty litter made Lowe so rich that he owned 22 houses and once bought an entire town in Michigan.

Cat poop money is good money.

Clumping litter was invented in 1984 by a Persian-cat-breeding biochemist named Thomas Nelson.

New, greener alternatives to the clay pellet litter include sawdust, newspaper, pine and beet pulp.

THE BEDROOM CLOSET
Shoes pg. 28
Silk pg. 32
Velcro pg. 36

Sports Bras
pg. 40
Safety Pins
pg. 44

SHOES
Jan Ernst Matzeliger was an intense guy.
I didn't come here to %&$#@ around.
Matzeliger was born the child of a black Surinamese slave and a Dutch engineer in Dutch Guiana in 1852.
At the age of 19, Matzeliger found work as an engineer on a Dutch East Indies Co. steamship and left to see the world.
This machine is... so...beautiful.
CHUG CHUG CHUG
Two years later, he disembarked in Philadelphia, looking for work as a machinist.
My résumé.
CHUG CHUG CHUG

Philadelphia was still very segregated, and Matzeliger's dark skin barred him from most shops. He could only find work as an apprentice shoemaker.
Ok.
A slight step down from huge awesome steamship.
Shoes at this time were a luxury item.
You're doing it wrong.
Only rich folks could afford to buy more than one pair of them.
I heard he owns *three* pairs!
They were costly because, while most of the process was automated, the upper part of the shoe had to be attached to the sole by hand.
Pfft. Silly humans. Getting in the way.
This was called "lasting," and it was the province of highly specialized (and slow-working) artisans.
Matzeliger calculated that automating this process would cut production costs in half.
He decided to invent a machine that would do the lasting work.
I didn't come here to %&$#@ around.
Might have mentioned that.
Jan Ernst Matzeliger went all in.

In 1883, after five long years, Matzeliger finally completed his automatic lasting machine.

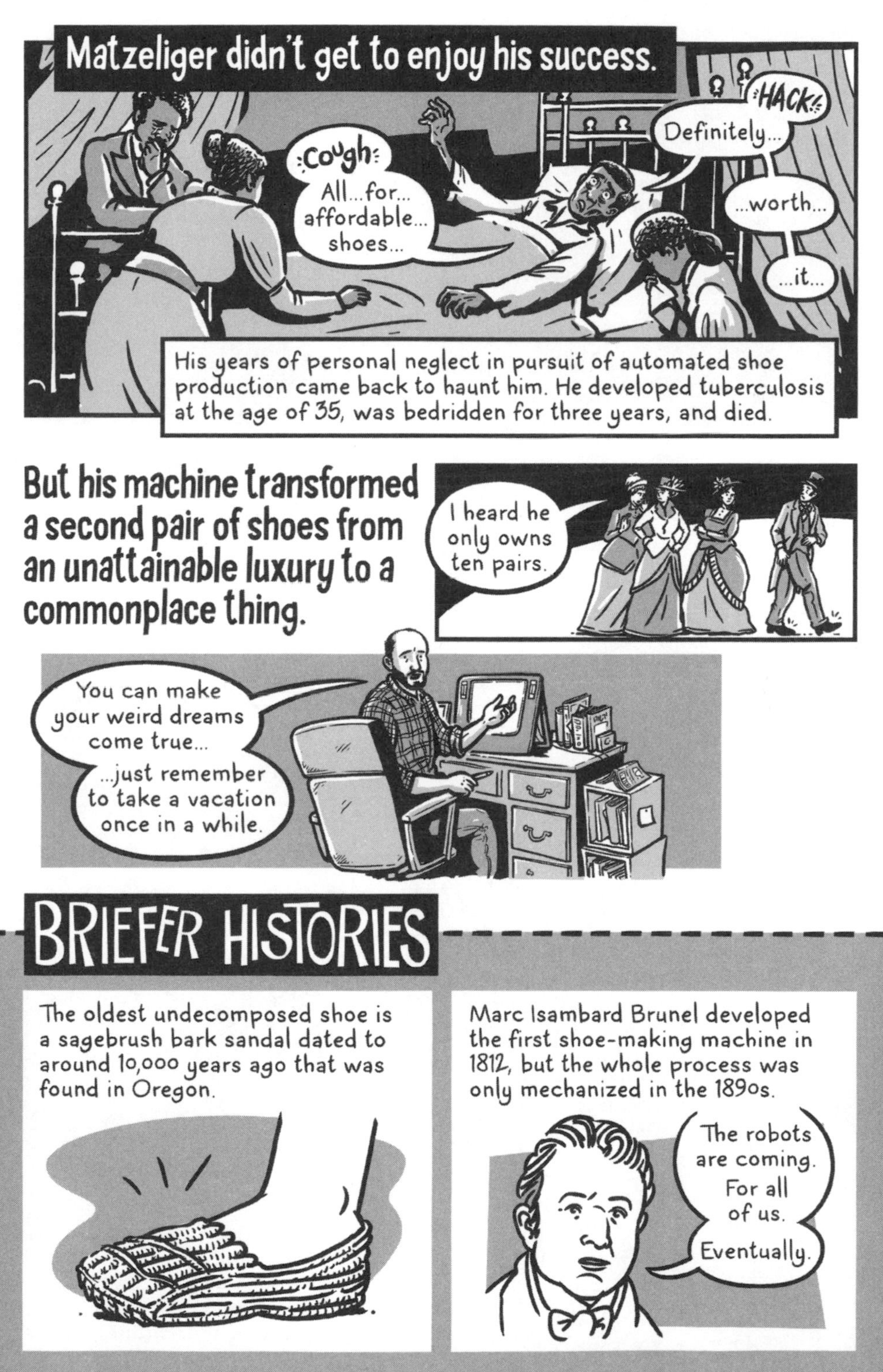
Matzeliger didn't get to enjoy his success.
Cough
All...for... affordable... shoes...
HACK!
Definitely...
...worth...
...it...
His years of personal neglect in pursuit of automated shoe production came back to haunt him. He developed tuberculosis at the age of 35, was bedridden for three years, and died.
But his machine transformed a second pair of shoes from an unattainable luxury to a commonplace thing.
I heard he only owns ten pairs.
You can make your weird dreams come true...
...just remember to take a vacation once in a while.
BRIEFER HISTORIES
The oldest undecomposed shoe is a sagebrush bark sandal dated to around 10,000 years ago that was found in Oregon.
Marc Isambard Brunel developed the first shoe-making machine in 1812, but the whole process was only mechanized in the 1890s.
The robots are coming.
For all of us.
Eventually.

SILK
Around 6,000 years ago, some folks living in northern China domesticated an insect and changed the world.
Now, sit!
I don't think it's working.
The insect was the Wild Silk Moth, and it ate the leaves of white mulberry trees.
Someone clever noticed that Silk Moth cocoons were awfully pretty, and began weaving them into fabrics.
OK, this is gonna sound crazy, but...
SHINE
Hmmm...
Someone even more clever realized that if you raised moths yourself, you wouldn't have to look for cocoons all the time.

Legend has it that Empress Leizu, wife of the Yellow Emperor of China, came up with the idea around 2500 BC when a cocoon fell into her cup of tea.
Uh... gross.
But the Yellow Emperor and his wife probably didn't really exist, so who knows?
Weiiird.
Dare you to put it in Mom's tea!
After generations of selective breeding, the Wild Silk Moth was transformed into the blind, flightless, modern domesticated silk moth.
Ta-da!
Over the course of a few days, adult silk moths lay thousands of eggs, then die.
Splort!
Munch Munch
Blarg
The eggs hatch into around 30,000 silkworms. These gobble up an absurd quantity of mulberry leaves, then make twelve pounds of silk cocoons.
People everywhere went NUTS for the stuff!
EUROPE
PERSIA
CHINA
EGYPT
ARABIA
INDIA
Great Ra! It's so shiny!!!!
Silk was traded far and wide, connecting the empires of antiquity and allowing the spread of ideas and cultures across the ancient world.
But all the silk came from only one place—China.

The silkworms and the mulberry leaves they ate were kept a secret by the Chinese to maintain a lock on production, create scarcity and drive up prices.
I mean, think about what would happen if this fell into the wrong hands?
Horrible silk shirts everywhere!
Silk was a rare enough sight that when Roman legions saw the silk banners of the Parthian Empire's army in 53 BC, they were shocked and fled in panic.
Jupiter! They're so shiny!!!!
Eventually, silkworms were smuggled out of China, but nobody figured out that they only ate mulberry leaves.
%$#@!
By the middle of the Han Dynasty, around 80 BC, silk had become so valuable that it was being used as a currency.
OK, one scarf and two handkerchiefs in change.
But the Chinese couldn't keep a lid on their insect farming forever and the secret began to trickle out.

BRIEFER HISTORIES

Silk trade gave its name to the Silk Road, which in turn inspired the name of a huge online drug marketplace that was busted by the FBI in 2013.

Getting silk out of cocoons ends up killing the silkworm. That's led to a boycott by People for the Ethical Treatment of Animals.

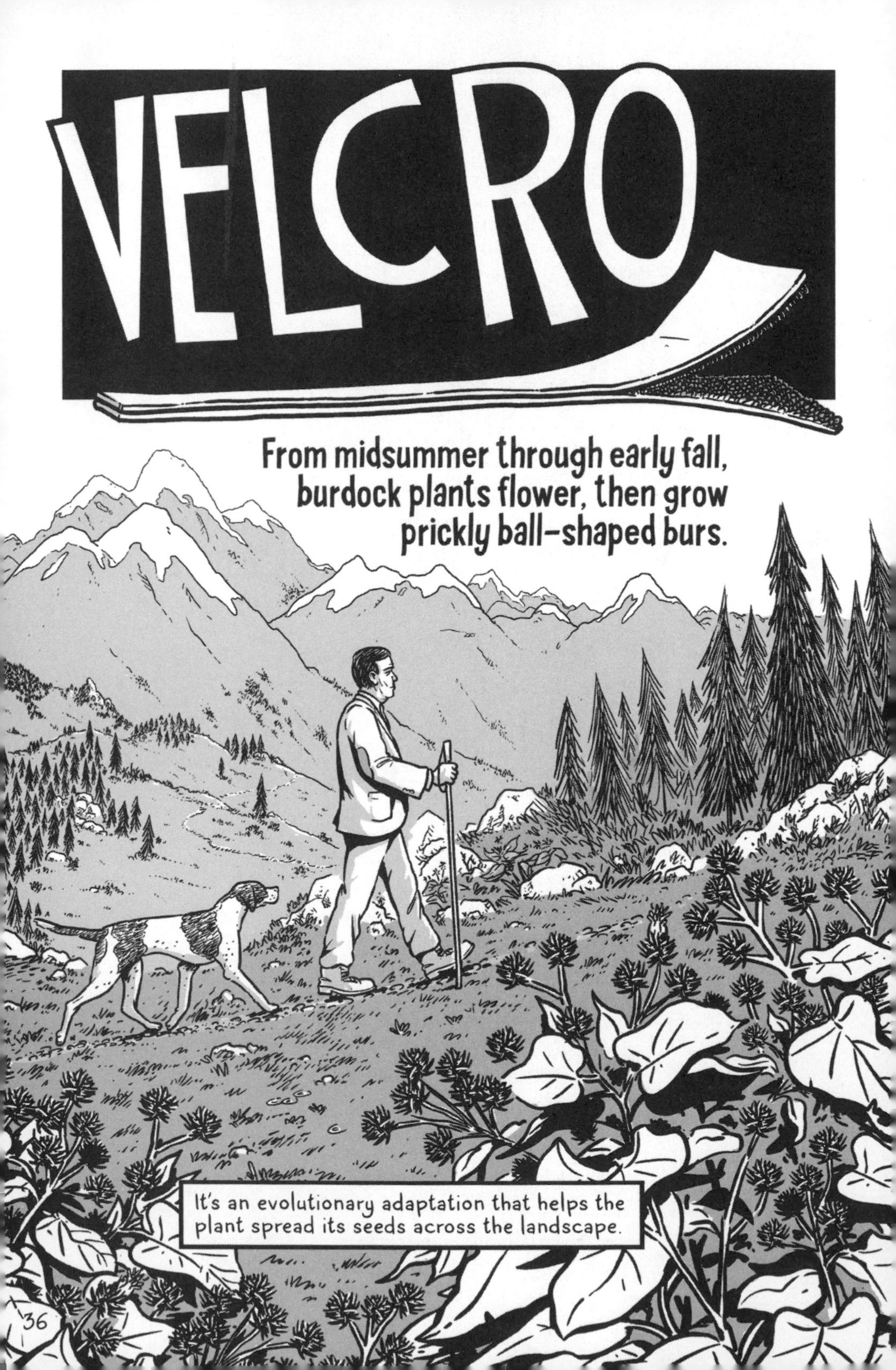
VELCRO
From midsummer through early fall, burdock plants flower, then grow prickly ball-shaped burs.
It's an evolutionary adaptation that helps the plant spread its seeds across the landscape.

De Mestral wondered at how the burs held on to his dog's fur so tightly...

...so he used a microscope to take a closer look.

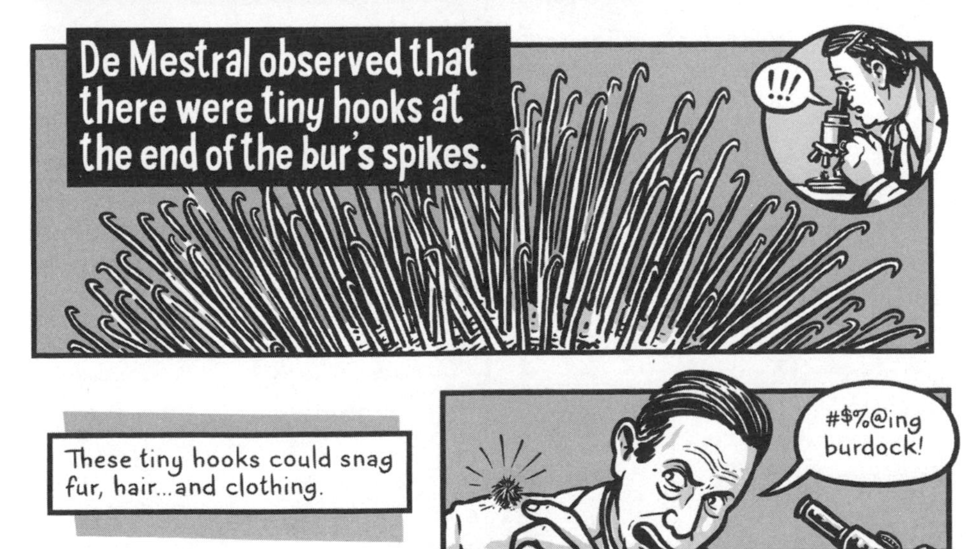

De Mestral remembered a few months earlier, when the zipper on his wife's dress had stuck.
Almost... hrf...got it!
Easy!
He realized that a fastener that mimicked the design of the burs wouldn't get stuck like that zipper had.
Ahem.
No.
That would be embarrassing.
arroo?
Then de Mestral spent almost a decade trying to make the idea stick.
Heh! I see what you did there....
Wait. A decade?
He made a prototype out of cotton, but it wore out quickly.
Maybe not the best way to keep pants up yet...
The strange new fastener was a bit of a tough sell, but de Mestral kept at it.
No, you see, it's like a zipper and a big spiky seed get together...
Snort!
Shh! Hee hee!
?
De Mestral tried a newly discovered material called "nylon" and found that it was perfect. After years of effort, he had a functional prototype for mass production.
Suck it, zippers.
It's go time.
CHUG CHUG CHUG
RRRip!

In 1951, de Mestral named his product Velcro, and patented it.
De Mestral hoped his invention would replace zippers entirely.
Velcro me up, dear.
That didn't quite work out.
But in 1966, NASA needed a way to secure things in zero gravity. They turned to the material that was inspired by burdock burs and dog fur.
RRRiiiip!!
And before he died, de Mestral got to see Velcro used in space.
#$%@ing burdock!
BRIEFER HISTORIES
Velcro's name comes from the French words for "velvet" (velours) and "hook" (crochet).
Velcro is a brand name (like Kleenex). The material is technically called a "loop-and-hook fastener."
But c'mon, who says that?

SPORTS BRAS
In 1970s America, jogging was king.
Well, jogging and pet rocks...
...it was a weird time.
An American gold medal win in the 1972 Olympic marathon kicked off a nationwide craze for running.
Jogging was popular with everyone, but many women found the trend uncomfortable.
Ow!
GASP
Jeez!
The mismatch between sports and breasts was nothing new.
The basic needs of women were overlooked?
Shocking!
In the 1890s, a bicycling sport corset was introduced. Results were mixed.

But as fashion moved away from restrictive garments, many women found that certain areas of their bodies had a bit too much freedom of movement.
God! Never thought I'd wish I had a corset.
WAP!
Some women took extreme measures.
A famous French athlete named Violette Morris underwent a double mastectomy because her breasts got in the way. She was also openly gay in the 1920s and eventually became a Nazi collaborator.
Ok, so my lifestyle might have been a bit unorthodox....
108
In 1975, Glamorise introduced the "Free Swing Tennis Bra."
FREE SWING TENNIS BRA
For the active woman who golfs, skis, bowls, skates, sails and cycles. On the go! Bra-Net action sides stretch with you. Terry cot lined Antron III© cups 32-36 A. 34-38 B, C, D.
$5
But it was basically just a regular bra with stretchy "action sides."
In 1977, Lisa Lindahl got a telephone call from her sister that changed women's sports.
Hello?
I need to talk to you about jogging!
Lindahl's sister had recently taken up jogging, but found its side effects painful and embarassing.
...just flopping all over the place!
Uh-huh.
Men have a jockstrap for their jiggly bits! Why don't women?

Lindahl was a jogger herself. She figured she was as qualified as anyone to tackle the problem.
...a jockstrap for women...
Sip
And another thi—
CLICK!
She partnered with two local theater costume designers. They tried a few prototypes but nothing worked.
Sorry, we're doing Camelot right now....
Then Lindahl's husband thought he'd have a laugh and grabbed two of his own jockstraps.
Here's your jock bra, ladies!
Gasp! That's it!
That's what?
The women sewed the jock straps together...
...and the sports bra was born!
Lindahl and her husband divorced soon after.
And I want my jockstraps back, too.
Lindahl went to grad school to get her life back on track. She and her friends started selling "Jogbras" to pay for it.
Imagine if running wasn't a painful and embarrassing experience!

BRIEFER HISTORIES

Hinda Miller, one of the costumers who cocreated the jogbra with Lindahl, went on to become a Vermont State Senator.

Polly Smith, the other designer, went on to make costumes for the Muppets.

Violette Morris also raced cars, owned a bike shop, and was assassinated by the French Resistance. Her nickname was "the Hyena."

Lindahl and her partners called the original Jogbra the "jockbra." It's now part of the collections of the Smithsonian Museum of American History in Washington, D.C.

Mosaics have been found in Sicily dated to 300 AD showing women exercising with bound breasts.

SAFETY PINS
Walter Hunt was a genius.
He invented a repeating rifle, knife sharpener, ice plough, fountain pen and ink stand.
But Hunt wasn't good at profiting from any of it.
In 1833, Hunt invented the sewing machine.
This will usher in the modern age of garment manufacturing!
Wait... oh...
But upon realizing that his contraption would put seamstresses out of work, he disowned it.
Oh, dear...

In 1849, Hunt found himself in a pickle.

He needed $15 to pay a debt. That was a big sum of money in 1849!

Hunt had gotten the wire into a shape that could be clasped and unclasped, with much less danger of a painful poke from the wire's sharp end.
And no possible fallout in the low-wage labor market! Hunt was elated.
Hey! The good guy's finally gonna win.
This time around, Hunt was determined to turn a profit.
Darn tootin'!
He patented his invention and sold it to W. R. Grace and Co. for the princely sum of $400!
Hunt paid off his $15 debt and pocketed the remaining $385...
Nice little profit from a bit of wire!

...only to watch W. R. Grace and Co. make millions from the pin design he'd sold them.

sigh I'm starting to sense a pattern.

Then other businessmen became obscenely wealthy manufacturing the sewing machines that Hunt (out of shame) had never patented.

Walter Hunt spent the last decade of his life embroiled in patent suits.

BRIEFER HISTORIES

Walter Hunt's grave in Brooklyn's Greenwood Cemetery sits in the shadow of the monument of Elias Howe, who got rich manufacturing Hunt's unpatented sewing machine.

Ancient Greeks used a similar design to the modern safety pin called a "fibula" to keep their togas from slipping.

A 21-foot-tall safety pin sculpture by Swedish American artist Claes Oldenburg stands in the courtyard of San Francisco's de Young Museum in Golden Gate Park.

Oldenburg has also sculpted a giant apple core, clothespin and garden hose.

According to Ukrainian superstition, pinning a safety pin to your clothes wards away evil spirits.

THE LIVING ROOM
Vacuum Cleaners pg. 50
Monopoly pg. 54
Dice pg. 58

Yo-yos
pg. 62
Slinkys
pg. 66
SKRITCH
SCRATCH

VACUUM CLEANERS
In 1901, Hubert C. Booth attended the London demonstration of a new cleaning machine.
What?
Perfectly normal way to spend a weekend!
"BLOW CLEANER"
The machine was a sort of automated broom and dustpan. It blew dirt away from itself into a collection box. It often missed.
Still *hack* a few bugs *cough* to work out.
VAROOM!!!
Later, at a restaurant, Booth couldn't get the machine off his mind.
Look, it's a passion.

Inspiration struck!
Booth put his napkin on a dusty chair, stuck his mouth on it, and sucked in air as hard as he could.
SUCK!
He almost choked to death.
When he recovered, he found there was a small ring of dust opposite where his mouth had been on his napkin. The machine should have *sucked* not *blown*!
I'll thank you for not making the obvious joke.
Just keeping it clean!
Really?
So Booth invented a vacuum cleaner.
"Keeping it clean."
God.
SCRIBBLE SCRIBBLE
To be more precise, Booth invented a gas-powered suction pump mounted on a carriage nicknamed "Puffing Billy."
CHUG!! CHUG! CHUG!!
The contraption was parked outside a house, and a hose was inserted through the window.
Woo!
VAROOM!!!
SNORT!
It reportedly terrified horses.

In 1902, Booth's vacuum snagged a big celebrity endorsement.
Puffing Billy was used to clean Westminster Abbey before the coronation of Edward VII.
A few centuries of hereditary rule can leave things a little dusty.
Booth sent 15 machines, which vacuumed up 26 tonnes of dust over the course of four weeks.
What's up with the horses?
CHUG!!
CHUG!
VAROOM!!
Vacuuming became a fad! British society women held tea parties to watch Puffing Billy clean their manors.
SNORT
Ah, the sound of progress!
CHUG!!
CHUG!!
NEIGH!
VAROOM!!
But in less than a decade, the march of technology caught up to Puffing Billy.
The VACUUM CLEANER COMPANY LTD.
DUSTLESS SYSTEM FOR CLEANING
CARPETS CURTAINS TAPESTRIES
About %$#& time!
In 1907, an asthmatic American janitor named James Spangler invented a portable electric vacuum cleaner using an old fan and a pillowcase.
Cough!
HACK!

William Henry Hoover bought Spangler's design in 1908 and marketed vacuum cleaners to the masses.
Not terrifying horses is actually a pretty good selling point.
-Introducing- The HOOVER ELECTRIC SUCTION SWEEPER
The HOOVER
People realized they didn't need to hire a giant, noisy carriage-drawn behemoth to clean their houses anymore. Puffing Billy was retired.
WHEEEEEZE....
The VACUUM CLEANER COMPANY LTD
CARPETS CURTAINS TAPESTRIES
CLOP CLOP
BRIEFER HISTORIES
Before becoming a janitor, James Spangler invented a grain harvester, hay rake and pedal wagon, but didn't make any money from them.
My name's Walter Hunt.
We've got a lot to talk about.
cough!
Earlier suction machines existed, but were powered by bellows or hand cranks. Booth's vacuum was the first one to be motorized.
In no way is this better than sweeping.
CRANK CRANK
The largest vacuum ever built was created in 2004 as part of a short-lived Discovery Channel TV show where engineers built large versions of everyday items.
Television gold!
The technology that powered early robotic vacuum cleaners like the Roomba was developed from U.S. military funding for mine-detecting machines.
WHIRRRR
Iraq war '03 to '04. You?
WHIRR

MONOPOLY
MONOPOLY
GO
The original version of Monopoly was the brainchild of a feminist radical anticapitalist woman from Washington, D.C., named Lizzie Magie.
Betchya didn't see that one coming.
Magie was a big fan of Henry George, a Progressive economist who wrote about the negative social impact that land monopoly had on rent.
Everyone needs a hobby.
PROGRESS AND POVERTY
Magie had an unorthodox idea for how to teach George's economic theories to the masses.
A board game about taxes and property.
Unbeatable combo, right?!

In 1903, she created "the Landlord's Game," which had players running around a board gobbling up property and charging rents.
And then the players realize the progressive power of a single land value tax!
SNORE
Magie patented the game and began distributing it herself.
Have you ever heard of Henry George's theories on...
The Landlord's Game
An economics professor and prominent socialist named Scott Nearing came across Magie's game and taught it to his classes.
SNORE
Better than a lecture.
His students at the University of Pennsylvania and the University of Toledo began making their own boards so they could play at home.
At some point, one of these boards was introduced to a Quaker congregation.
The Quakers made their own boards, changed the rules a bit and renamed the game Monopoly.
Better than a sermon.
SNORE
Society of Friends
Cheat and I'll bash your head in.
Monopoly spread through Quaker groups throughout the Northeast and Midwest.

Homemade copies of the game continued to spread.

Darrow sold his version of the game to Parker Brothers.
OK... so you try to get rich and crush your enemies.
MONOPOLY
Parker Brothers
Americans were eager to get the grinding Great Depression off their minds and play at being wealthy. People bought the game in droves.
GO
When Magie began to make noise about the rip-off, Parker Brothers quietly bought out her patents and pretended nothing ever happened.
Never speak of this again.
BRIEFER HISTORIES
In Nazi Germany, Monopoly was denounced as "Jewish-speculative" propaganda and banned.
The Führer prefers Yahtzee.
Magie invented other board games, too. She even guilted Parker Brothers into publishing them!
Least they could #$%ing do...
Bargain Day
The other games didn't catch on.
In 1904, Magie held a stunt auction where she "sold herself as a slave" to protest women's low wages.
Check out my board game, too!
FOR SALE
Magie's story was only uncovered in 1973 after Parker Brothers sued to stop a satirical game called Anti-Monopoly.
ANTI-MONOPOLY
The "Bust-the-Trust!" Game
Parker Brothers lost the suit.

DICE
All around the world, archaeological digs have uncovered small objects that were probably used to either divine fortunes or bet away one's worldly possessions.
These often took the form of the ankle bones of an unfortunate animal.
Cheat and I'll bash your head in.
Someone had the idea to file the edges of the bones down in order to make a cube.
I mean, the head-bashing really wears on you after a while....
file file file file file file file file file file file
That evened the odds of any of the six sides turning up when rolled.

The earliest 6-sided die that we'd recognize was dug up in a city on the Eastern border of present-day Iran.
The site is called the Burnt City. It burned down three times before its inhabitants finally abandoned it in 1,800 BC.
It'll never happen again, you said.
Two times is unlikely enough, you said.
Two dice were discovered there in a 5,000-year-old proto-backgammon set.
The Burnt City wasn't only home to gamers – digging there also unearthed the oldest artificial human eye.
And clues that the city's residents wove baskets using their teeth.
I'll admit it's not the most pleasant technique....
There's a graveyard with tens of thousands of tombs.
But we have no real idea who lived in the Burnt City, where they came from, or where they went.

But if humans can take something relatively simple and then make it complicated and kind of bizarre — they will.

But its shape is immediately familiar to any geek who was a teenager after the 1970s.
I cast magic missile at the orc priest!
Ooh. Critical hit.
The 20-sided die is associated with Dungeons and Dragons, the mother of all role-playing games.
D&D's complex rules demanded more varieties of dice in gameplay.
From nerd to nerd, the game and its weird dice spread around the world.
And when the ruins of our civilization are dug up, it's a good bet that they'll puzzle over the small objects they find, too.
My guess is either for telling fortunes or betting away worldly possessions.
Beep?
BRIEFER HISTORIES
Loaded dice rigged to roll lucky numbers were found in the ruins of Pompeii.
RUMBLE
Ulp... not lucky enough!
Board games are even older than dice. Ancient Egyptians, Chinese, Aztecs, and Mesopotamians all played different ones.
Cheat and I'll bash your head in.

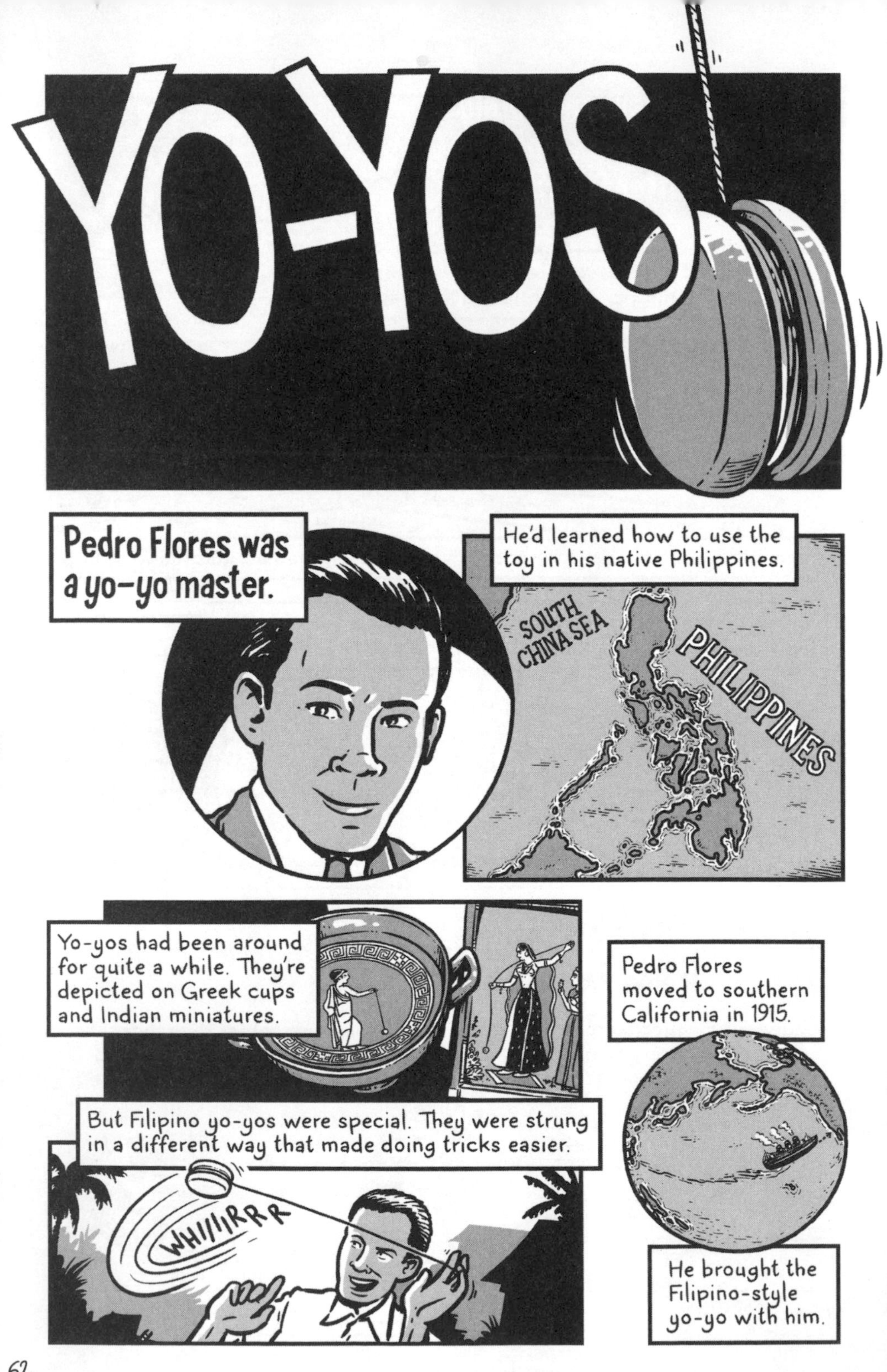
YO-YOS
Pedro Flores was a yo-yo master.
He'd learned how to use the toy in his native Philippines.
SOUTH CHINA SEA
PHILIPPINES
Yo-yos had been around for quite a while. They're depicted on Greek cups and Indian miniatures.
Pedro Flores moved to southern California in 1915.
But Filipino yo-yos were special. They were strung in a different way that made doing tricks easier.
WHIIIIRRR
He brought the Filipino-style yo-yo with him.

His skills proved an immediate hit with the locals.
Marge!
You gotta see this!
Holy cannoli!
I smell a lucrative business opportunity....
Good god!
Yowza!
WHIRRRRR
So in the late 1920s, Flores acquired investors, set up a factory and began mass-producing yo-yos.
TOY STORE
NEW
FLORES PAT. PEND. YO-YO
Flores wasn't the only one to notice the yo-yo's capacity for drawing a crowd.
A San Francisco ice cream salesman named Donald Duncan saw a kid playing with a yo-yo, surrounded by a crowd of onlookers.
I smell a lucrative business opportunity....
Duncan decided to get in on the game.

Duncan bought Flores's patents and factory for $250,000.
Don't spend it all in one place.
That was a huge sum at the time!
$ $ $
Duncan then traveled to Filipino neighborhoods in southern California, where he found young local yo-yo champs and hired them by the dozen.
Hey! You guys!!
I have a lucrative business opportunity!
He paid the yo-yo masters to fan out across America, dazzle neighborhood children with tricks, and then direct them to a nearby toy store for purchase.
TOY STORE
WHIRR!
It worked like gangbusters.

The yo-yo became a nationwide craze.
Relentless TV advertising after World War II pushed Duncan yo-yos to new heights.
By the 1960s, Duncan's factory was making 60,000 yo-yos a day.
CHOP! WHIRR
THK! THK!
Four out of five yo-yos in America were Duncan brand.
But in 1965, Duncan's company lost copyright on the word "yo-yo," when a judge noted it was the common term for the toy in the Philippines.
I'll show you &%$# common terms.
Then the yo-yo craze dried up, leaving Duncan's company with huge debts and a giant, useless factory.
CHOP thk whrr thk sputter...
DUNCAN
Undone by the yo-yo's popularity, the company went bankrupt.
BRIEFER HISTORIES
The most expensive yo-yo was sold in 1992 for $16,000. It was signed by Richard Nixon, who was notoriously terrible at yo-yo.
$ $ $
Both Napoleon and Lord Wellington and were yo-yo fans. Alas, history does not record a yo-yo battle between them at Waterloo.

SLINKYS
In 1943, Richard T. James was a straitlaced naval engineer working on securing instruments against the rocking of the ocean.
BUMP
A clumsy elbow or a lurch of the boat knocked a large torsion spring off his desk.
Richard was flabbergasted at the strange way it moved.
Gasp! Look at it walk!
SPROING!
He brought the spring home to show his wife, Betty.
Oh, boy!
Have I got something to tell you about!
SLAM!

Betty wasn't sold on it.
A spring, Richard?
Really?
Richard persisted. He experimented with different types of steel wire.
Richard? It's late.
PROD!
Wobble
Finally, he created a prototype that could smoothly walk down stairs.
Richard showed it to the neighborhood kids. When Betty saw their reaction, she became a true believer.
Gasp! Look at it walk!
Betty dug through the dictionary to find a better name for the toy. She settled on "slinky," because it reminded her of how the toy would walk down a slope.
I dunno—I still kind of like "torsion spring."
Sounds fun-loving!
DICTIONARY
DICTIONARY
The couple started a company, produced 400 Slinky toys and tried selling them to stores.
A spring, Mr. and Mrs. James?
Really?
A spring that walks!
They didn't have much success.
In November of 1945, they got their big break. Gimbels department store in Philadelphia let them set up an inclined plane at the end of a counter to demonstrate.
Shhh!
"Slinky"
$1.00
Betty and a friend came to boost sales by pretending to be customers.

The Slinkys sold out in a 90-minute consumer frenzy.

Betty wasn't sold on it.
Um. Anyone?
So Richard moved to Bolivia by himself, leaving Betty with six kids, a divorce, and a broke Slinky company.
AIR LINE
Well... look at him walk.
Betty spent the next 38 years building the Slinky into one of the most iconic toys of the 20th century and retired very wealthy in 1998 at the age of 80.
I mean really, the damn thing practically sells itself.
Gasp!
Gasp!
Gasp!
Richard died in Bolivia in 1974.
BRIEFER HISTORIES
In 1952, a woman named Helen Malsted created the Slinky Dog. Betty immediately licensed it.
Gasp!
John Cage composed a musical piece in 1959 involving a Slinky, a broom and a cage of canaries.
The evangelical translators to whom James gave his money claim to have translated the bible into over 2,800 languages.
The Slinky is the official toy of the state of Pennsylvania.
In case you were wondering.

THE KITCHEN
Tupperware pg. 72
Microwave Ovens pg. 76
SIZZLE

Coffee Filters
pg. 80
Trash Cans
pg. 84

TUPPERWARE
At the 1954 Tupperware jubilee, hundreds of women dug through a Florida swamp for buried treasure.
A mink coat!
A gold watch!!
A new car!?!?
The women were Tupperware dealers and a trip to the jubilee was a reward for good sales at house parties.
Listen, let me tell you about plastics.
All told, the total value of the buried prizes was over $70,000.
A NEW #&@$ING CAR!?!?!?
The jubilee — and the huge Tupperware party sales it celebrated — were the brainchild of the container company's vice president, Brownie Wise.
A new #&@$ing car.

Tupperware itself was created from an industrial byproduct called polyethylene slag.
In 1938, a man named Earl Tupper figured out a way to refine the slag into a flexible, durable plastic material.
This industrial byproduct has potential!
Eight years later, he figured out how to make it into a container...
...which Tupper then named after himself.
Polyethyleneware was a pretty close second, though.
BURP
Tupperware didn't sell well.
Ok, what if it was called "Slagware?"
GENER
STO
People were used to glass or ceramic containers, and it wasn't immediately apparent how to close Tupperware lids.
$&#@!
GASP!
Then Brownie Wise came along and everything changed.
Wise sold cleaning products at home parties she organized. She was great at her job, but her boss told her he'd never promote a woman to an executive level.
Look, I'd prefer to cling to my baseless notions of female inferiority.
I'm sure you understand.
Brownie Wise looked for greener pastures. She found Tupperware.

In 1949, Wise started selling Tupperware at house parties, where she could demonstrate how the lids worked.

Wise was an insanely good salesperson. Within a year, she'd cleared $150,000.

Tupper abruptly fired Wise in 1958.

Tupper erased Wise from Tupperware company history, then buried hundreds of copies of her memoir in a pit.
It's #&@$ing TUPPERWARE!
GET IT?
Whoa!
Then he evicted her from her company-owned house. Wise sued, but she only received a year's pay as a settlement.
TUPPERWARE!!
Jeez!
Then Tupper sold the company for $16 million and bought an entire island off the coast of Costa Rica.
TUPPERWARE!!!
Um... and that's really how it actually happened.
I'm so, so sorry.
Wise started a few other sales companies, but nothing took off. Eventually she became a potter.
BRIEFER HISTORIES
Tupperware wasn't Earl Tupper's only invention. He also came up with an ice cream cone that didn't drip and a boat powered by fish.
Tuppercone and Tupperboat!!!

Selling Tupperware allowed many American women to earn enough to support themselves despite the rigid gender roles of the 1950s.

MICROWAVE OVENS
It was 1945, the tail end of World War II, and Percy Spencer, a radar engineer at Raytheon, was hungry.
RUMBLE
Spencer reached into his pocket for a chocolate bar...
Well, I'll be!
...only to find that the bar had melted!
Spencer's suspicion fell on the magnetron he'd been standing next to that had been steadily pumping out microwave radiation.

Raytheon had been supplying cavity magnetrons to the American military for use in radar systems.
They gave a crucial advantage to U.S. troops over their German and Japanese enemies.
Spencer, however, was intrigued by their newly discovered ability to melt food.
Spencer grabbed a handful of corn kernels and scattered them in front of the machine.
All in the spirit of scientific inquiry!
Yowza!
The results were dramatic.
POP! POP! POP! POP! POP! POP! POP! POP! POP! POP! POP!

More data was needed!
Spencer cut a hole in a tea kettle, put an egg inside, and called a colleague over to watch.
Why is there popcorn everywhere?
Shh! We are EXPERIMENTING!
HRRRMMMM
The results were dramatic.
The commercial applications are obvious!
Splort!
Spencer was put in charge of developing commercial microwave technology for Raytheon.
OK, so pudding also explodes.
Within 2 years a prototype was built.

And Percy Spencer kept blowing up food.
All in the spirit of scientific enquiry!
His name is on the patents for everything from popcorn bags to a method of microwave-broiling lobsters.
BRIEFER HISTORIES
Spencer received only a $2 gratuity from Raytheon for the invention of the microwave.
But I got to blow up lobsters.
Totally worth it!
Spencer ended up holding over 300 patents, despite never having graduated from high school.
U.S. PATENT
PATENT
The first microwave oven available to the public weighed 750 pounds and stood just under 6 feet tall.
Microwaves soon became smaller. By the end of the 1970s, they outsold conventional ovens.

COFFEE FILTERS
Coffee-brewing methods in 1908 left a lot to be desired....
A sludge of grounds was usually left in the bottom of the cup.
Melitta Bentz, a 34-year-old German housewife, was tired of crappy coffee.
Need... better... coffee...
...Mom?

Bentz tore a page of blotter paper out of her sons' school notebook...
More important than homework.
But my math homewo—
rrrRRIP!!
BANG! BANG! BANG!
...mutter mutter...
...then punched some holes in the bottom of a brass pot.
Bentz stuffed the blotter paper in the pot, added coffee grounds and poured hot water through it.
...mutter mutter...
1
The coffee brewed in the pot.
2
But only the liquid passed through the blotter paper.
3
The brew was less bitter and grounds-free.
Better coffee!
Coffee filtration was born!

Soon, her filters were selling in the thousands.

The war ended with German defeat. Bentz bounced back and expanded production.
Better coffee?
Nazis took over Germany in 1933. Six years later, war exploded again.
Bentz's factories were forced to stop making filters and pump out goods for the army.
The Führer prefers tea.
When the Nazis were defeated, Bentz's factories were seized by Allied forces.
CLOSED
Oh, come on!
But Bentz kept making filters, and slowly rebuilt her company as Germany's economy ground back to life.
Better... coffeee...
Seven years after Bentz's death in 1950, her company finally got the factories back.
BRIEFER HISTORIES
In 1674 in England, the Women's Petition Against Coffee claimed the drink rendered husbands impotent.
Definitely the coffee.
ahem
Definitely.
The most expensive coffee in the world is eaten, then pooped out, by a civet, which looks a little like a weird cat-rat-raccoon hybrid.

TRASH CANS
Lillian Moller Gilbreth was one hell of an efficient woman. She had to be.
Dear, did you look at the draft we got back from the editor?
Next on the list.
After five pairs of diapers.
By 1922, she was a successful author, a respected psychologist, and engineer, held a master's degree and a Ph.D. and was the mother of 11 children.
Lillian and her husband Frank were pioneers in using motion study to increase efficiency in industrial settings.
No, faster and better. At the same time.
And right now.

The Gilbreths experimented with their own children to troubleshoot many of the techniques they developed.
OK. If we have the soap already on the washcloth, we save four seconds.
But only if they don't cry...
Lillian and Frank founded a management consulting firm called Gilbreth Inc.
Oh, your employees are troublesome?
Have you met a toddler?
Their methods caught on.
Then Frank died suddenly of a heart attack in 1924.
A devastated Lillian found herself responsible for 11 children and a management consulting firm.
Only one way to ameliorate the situation:
Even more efficiency.
SNIFF!
AH-HN...
SNRF!
SNIFF!
She didn't slow down for a minute.
Lillian was contracted to consult for big companies like Macy's and Johnson & Johnson.
No, faster and better. At the same time.
And right now.
In 1929, Lillian turned her laser focus on the kitchen.

DATA FOR
TO MEASURE
NAME:
ADDRESS:
DATE:
1. HEIGHT ______ inches.
4. HIP MEASUREMENT ______ inches.
10. The present height of chair ______ inches.
BROOKLYN BOROUGH GAS COMPANY
Lillian interviewed thousands of women to determine the proper height for stoves, sinks, and cabinets.
Counter
Counter
Sink
Work triangle
Refrigerator
Counter
Stove
She pioneered refrigerator-door shelving and new, more efficient kitchen layouts.
By Lillian's measurements, the number of physical steps to prepare a shortcake was cut by more than two thirds!
THE KITCHEN PRACTICAL
I AM A GOD MADE FLESH!
One of the most enduring inventions from Lillian's "Kitchen Practical" was the pedal-operated trash can.
In contrast to holding trash while bending down to wrestle open the filthy trash can lid with one hand...
Ew.
...Lillian's design was a simple, hygienic, and elegant solution.
And efficient!
It caught on, and foot-pedal trash cans spread across the kitchens of America.

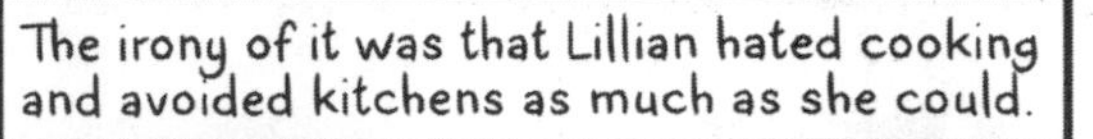

She designed sanitary napkins and desks, became a professor, was on the board of directors of the Girl Scouts, advised the U.S. Navy during WWII and died aged 93 in 1972.

BRIEFER HISTORIES

Lillian almost had *two* Ph.D.s, but was denied a degree for breaking residency. She published her thesis as a bestselling book.

Cheaper by the Dozen was adapted into a 2003 Steve Martin movie. Aside from the title, it had little resemblance to the Gilbreths' story.

Lillian also served in the administrations of Presidents Hoover and Truman and sat on the Chemical Warfare Board.

THE COFFEE SHOP
Tea
pg. 90
Artificial
Sweeteners
pg. 94

EXIT
Cinnamon
pg. 98
AMON
Coffee Beans
pg. 102

TEA
Tea drinking probably reached its apotheosis during the late Sengoku period, in 16th-century Japan.
HACK!
SLASH!
STAB!
The country was at the tail end of two centuries of tumult and war. It was as good a time as any to cultivate insanely complex calming rituals.
Tea had crossed the sea from China in the 9th century, but the Sengoku-era ceremony took it to a whole new level.
Pfft. Amateurs.
Show offs.
For one thing, it lasted for *hours*.
I'm getting a cramp!
Shh!
Princes hired their own tea advisors to help them choose from over 60 varieties.
Uh...this one is...tangy?

The ceremony involved philosophical discussion...
What is the true nature of the universe?
Tough one.
...precisely prescribed motions...
...and ritualized small talk.
How're the kids?
You're doing it wrong.
When Dutch traders first encountered the tea ceremony in Japan in 1610, they were taken aback.
And they didn't get up to pee once!
It's insane!
The Dutch began importing tea (mostly from China), but replaced the zen rituals with pleasantries and small tea cakes.
What is the true nature of the universe?
You're doing it wrong.
Tea only reached England in the 1660s after Charles II married a Portuguese princess with a penchant for the stuff.
Have you noticed how we all go bonkers for anything our royal family does?
Siiip
Siiip
But when the English fell for tea, they fell hard.
CHATTER CHATTER
Blimey!

The British East India Company placed its first order for tea from China in 1664.
For some reason, I have a bad feeling about this....
Within 40 years, three-quarters of all trade between East Asia and Europe was in tea.
Even the porcelain cups in which tea was served were a hugely profitable import.
A massive trade imbalance began to develop.
You paid them how much for it?!
British silver reserves were being shipped off to China in exchange for tea.
Tea literally grows on trees. This is amazing.
But the Chinese had no interest in European imports.
%$#&ing cheap foreign knockoffs!
The British had a problem. Their solution was opium.
Yes, yes, this seems to be the ticket.

BRIEFER HISTORIES

A Chinese legend about the origins of tea claims the first plants sprouted from the cut-off eyelids of a famous bearded, blue-eyed Buddhist monk.

Sen no Rikyū developed the complex Japanese tea ceremony in the 1580s, then ran afoul of the Shogun and was ordered to ritually kill himself. The last thing he did was serve tea.

ARTIFICIAL SWEETENERS
In 1879, German chemist Constantin Fahlberg sat down to eat without washing his hands.
He was astonished to find that everything he touched tasted outrageously sweet!
Mein Gott!
It is like candy!!!
Fahlberg realized that one of the coal tar substances he'd been testing had gotten stuck on his fingers...
...but he didn't know which substance was the sweet one!

So Fahlberg drank the contents of every beaker in his lab.
glug
This coal tar substance is delicious!
Luckily for him, none of them were toxic.
One beaker contained a new discovery, the first artificial sweetener, SACCHARIN.
O
N
S
O
O
It's now used in Sweet'N Low.
For obvious reasons, chemists aren't supposed to taste their experiments. You'd think that something like this would only happen once.
But it happened three more times!
burp Oh ja?
Oh ja.

The Second Time
In 1937, Michael Sveda discovered **cyclamate** (Sweet 'N Low*) when he noticed that his cigarette tasted sweet after a day working in the lab.
Na
N
S
O
This cigarette is *delicious!*
*In Canada. Cyclamate is banned in the United States.
This paper is *delicious!*
The Third Time
In 1965, chemist James Schlatter licked his finger to pick up a piece of paper, and discovered **aspartame** (NutraSweet, Equal).
O
N
The Fourth Time
In 1976, Leslie Hough told Shashikant Phadnis to test a chlorinated sugar compound. Phadnis thought Hough had said "taste it," so he did. They discovered **sucralose** (Splenda).
This chlorinated sugar compound is *delicious!*
O
Cl

BRIEFER HISTORIES

In ancient times, a treated lead compound called "Salt of Saturn" was used as a sugar substitute to sweeten foods. It was toxic and killed a pope!

Saccharin is 300 times as sweet as sugar. Cyclamate is 30 times as sweet, aspartame is 160 times as sweet, and sucralose is **600 times as sweet!**

Saccharin was almost banned in 1977 when lab rats got cancer from it, but it turned out that the cancer was unique to the rats and didn't affect humans.

Benjamin Eisenstadt invented the paper packet that artificial sweeteners are served in...but he forgot to patent the design!

CINNAMON
Cinnamon was worth more than 14 times its weight in silver in 1st-century Rome.
But the Arab traders who brought the cinnamon sticks to ancient Europe had a pretty good reason for the high price.
GIANT BIRDS!!!
What? Holy cow!
Seriously.
We get this stuff from big... freaky... birds.
Yow!

The traders convinced their customers that the giant flesh-eating birds used cinnamon sticks to build their nests on top of huge rocks.
The Arabs would leave out chunks of dead oxen, which the hungry birds would carry back to their nests.
The meat chunks knocked cinnamon sticks loose.
SPLORAT!
And the Arabs gathered the spices where they landed on the ground.

The prospect of giant, carnivorous birds was apparently freaky enough to leave Arabs in control of the cinnamon trade for 3,000 years.
Man! Those guys are badass!
This came to an end in 1505, when Portuguese ships arrived in Sri Lanka — and realized they'd found where cinnamon actually comes from.
Wait, where are the terrifying giant birds?
Shhh!
The Portuguese decided to cut out the middlemen.
Mmm. Smells like the holidays.
They deposed the local kings, occupied swaths of the island...

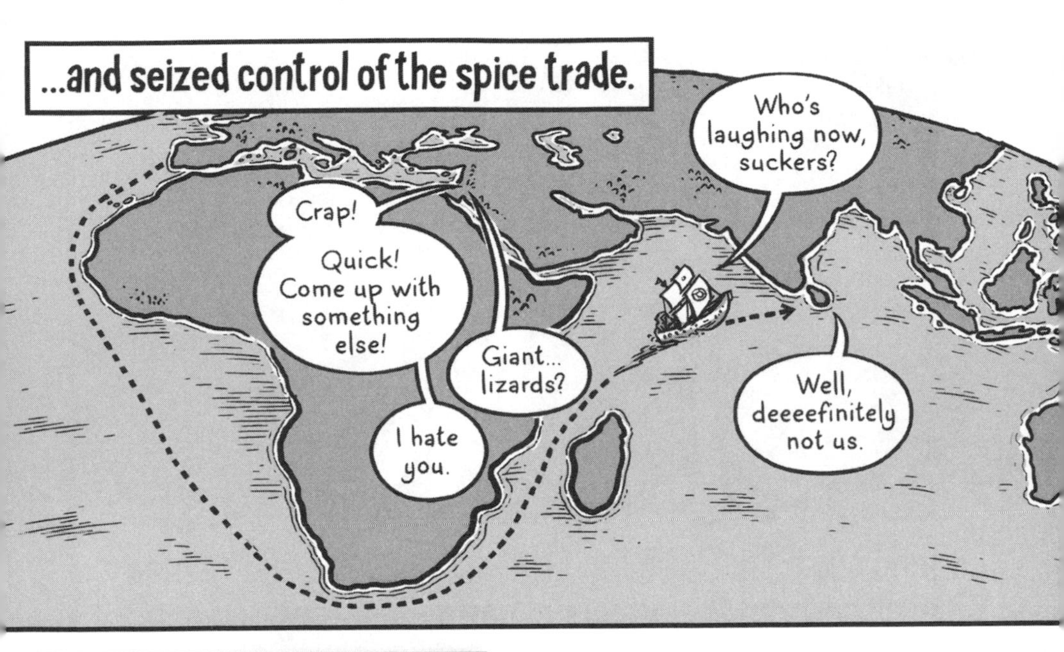

BRIEFER HISTORIES

The Roman Emperor Nero burned a whole year's worth of cinnamon on the funeral pyre for his wife... whom he'd kicked to death.

In Exodus, God commands Moses to mix cinnamon, myrrh, cassia, and calamus with olive oil and pour it on the ark of the covenant.

Modern cinnamon is almost all made from Chinese cassia trees, rather than the "true cinnamon" of Sri Lanka. Mexico buys most of it to make delicious chocolate.

Swallowing a tablespoon of ground cinnamon, then choking on it was a brief YouTube craze in 2012.

COFFEE BEANS

Legend has it that a young Ethiopian goat herder named Kaldi discovered the effects of coffee beans by accident.

Kaldi noticed his goats acting frisky after eating the berries from a certain bush, so he tried a few himself.

The berries hit Kaldi like a freight train.

He brought the beans to the attention of a local monastery, where the abbot declared them the work of the devil and ordered them burned.
GET THEE BEHIND ME, SATAN'S BERRIES!
But the roasting aroma was so delicious that the monks pulled the beans from the fire!
Ow! %$#!
Kaldi and his goats could be a myth, but coffee did travel to the Arab world through Yemen's trade with Ethiopia.
Trust me on this one....
SNRF
The bitter brew spread with Islam, but production was kept secret.
Big freaky bir—
Shhh!
The Dutch nicked some plants in 1616. Coffee production took off.
Trust me on this one....
In 1723, a French naval officer stole a coffee plant from the garden of King Louis XIV of France and smuggled it across the Atlantic into Martinique.
And I only get one tiny panel about that?

She gave Palheta a bouquet with coffee seedlings hidden inside.

He brought the seedlings back to Brazil and planted them in the state of Pará in 1727.

Palheta's contraband became the foundation of the largest coffee-growing empire in world history.
COFFEE of the DAY
BRAZILIAN DARK ROAST
Refills are a quarter.
At its peak in the 1920s, Brazil supplied almost 80% of the world's coffee. It's still by far the world's foremost producer.
BRIEFER HISTORIES
Around 1600, Pope Clement was asked to condemn coffee as "Satan's trap to catch Christian souls," but he tried a drink and thought it was too delicious to be evil.
gurgle
gurgle
In 1670, a man named Baba Budan smuggled seven coffee plants from Yemen to India. He's revered as a Sufi saint.
In 1718, the Irish Parliament banned mixing coffee beans with sheep droppings.
Don't blame me for crappy coffee!
That French officer from 1723 had to split his water rations with the coffee plant and was almost kidnapped by pirates.
Oh, great. I get a footnote, too.
PIRATES!

THE OFFICE
Paper Clips
pg. 108
Ballpoint Pens
pg. 112
Paper
pg. 116

Pencils
pg. 120
Post-it Notes
pg. 124
Stamps
pg. 128

PAPER CLIPS
The nation of Norway has a thing for paper clips.
There's a 26-foot-tall statue of one in Oslo.
Say clippy!
In 1999, the country featured them on a commemorative stamp.
Norge 4.00
They even served as a symbol of Norway's resistance to the Nazi occupation during WWII.
Well, this went from whimsical to dark real fast.

Symbols of the Norwegian nation, like images of the flag or exiled king, were banned by the Nazis.
So Norwegian students wore paper clips on their lapels to show their collective defiance of the occupation.
Office supplies of resistance!
The Nazis banned paper clips, too.
Were you expecting Nazis not to be @%$#heads?
All of this love for paper clips comes from their status as an iconic Norwegian invention.
Probably the first thing that springs to mind, right?
Johan Vaaler dealt with a lot of paper working as a clerk in Oslo in 1899.
There's gotta be a better system!
So Vaaler invented and patented a clip to keep everything a bit more organized.
Norwegian ingenuity at work!

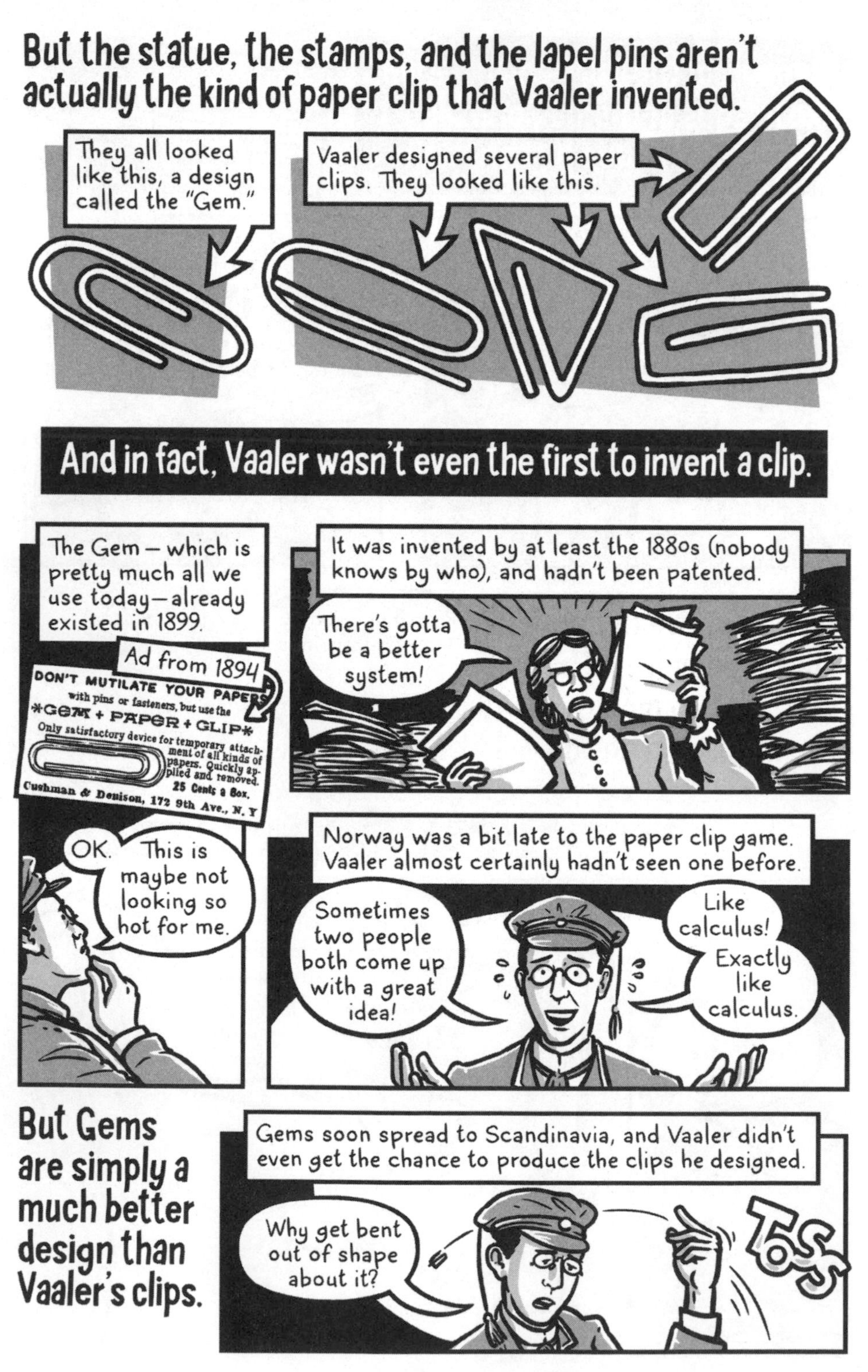
But the statue, the stamps, and the lapel pins aren't actually the kind of paper clip that Vaaler invented.
They all looked like this, a design called the "Gem."
Vaaler designed several paper clips. They looked like this.
And in fact, Vaaler wasn't even the first to invent a clip.
The Gem — which is pretty much all we use today — already existed in 1899.
Ad from 1894
DON'T MUTILATE YOUR PAPERS with pins or fasteners, but use the *GEM + PAPER + CLIP* Only satisfactory device for temporary attachment of all kinds of papers. Quickly applied and removed. 25 Cents a Box. Cushman & Denison, 172 9th Ave., N. Y
It was invented by at least the 1880s (nobody knows by who), and hadn't been patented.
There's gotta be a better system!
OK.
This is maybe not looking so hot for me.
Norway was a bit late to the paper clip game. Vaaler almost certainly hadn't seen one before.
Sometimes two people both come up with a great idea!
Like calculus!
Exactly like calculus.
But Gems are simply a much better design than Vaaler's clips.
Gems soon spread to Scandinavia, and Vaaler didn't even get the chance to produce the clips he designed.
Why get bent out of shape about it?
TOSS

Vaaler's story caught on and the paper clip became a national symbol, facts be damned.

BRIEFER HISTORIES

Not to rub salt in it, but the Nazi-defying students probably weren't referencing Vaaler. His story only became known after WWII.

In 2004, UC Davis student Dan Meyer made a 5,340-foot-long paper clip chain in 24 hours for a Guinness World Record. Eight years later, another Davis student named Jester Jersey spent a year training to beat it.

Jester's 2012 victory claim remains unrecognized.

The Gem wasn't the first paper clip either. The first was invented in 1867, by Samuel Fay. It didn't work too well.

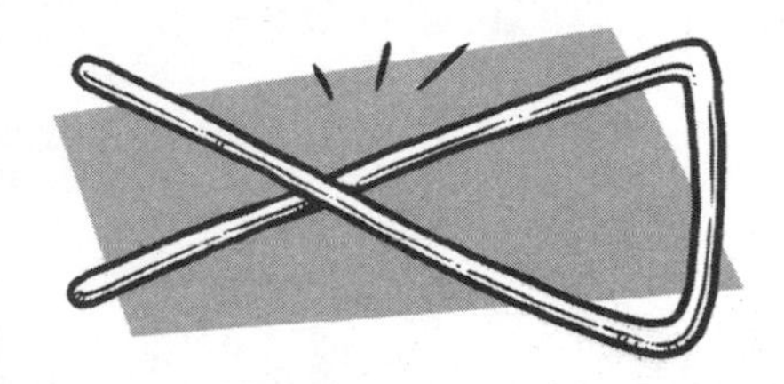

Dozens of other paper clip designs were invented and patented after it, but none took off like the Gem.

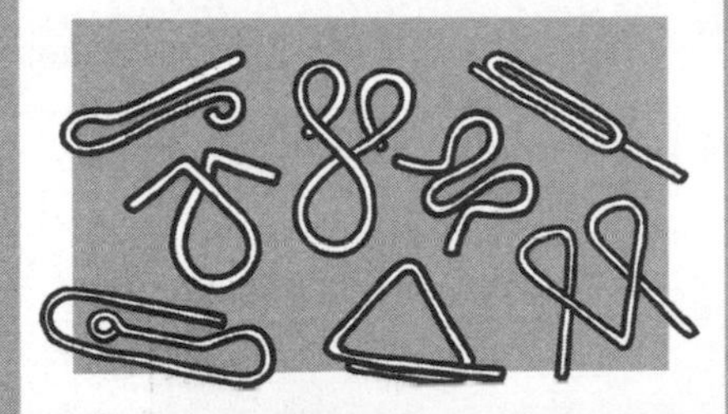

BALLPOINT PENS
On October 29, 1945, a crowd of over 5,000 people mobbed the Gimbels department store in New York City. They wanted one thing.
GIMBELS
BALLPOINT PENS!!!
Gimbels sold out of 10,000 pens in a single day... at $12.50 a pop!
OFFICE
Ballpoint pens!
The ballpoint pen craze was off with a bang.
The pens had been invented seven years earlier, in a place far, far away from New York City.
Hungary!
László Bíró, a Jewish Hungarian journalist living in Budapest, noticed that newspaper ink dried much faster than his fountain pens.
Good god.
NEWS
Germany invades Sudetenland!
This ink is amazing!

Bíró designed a pen that could deploy newspaper ink using a rotating ball mechanism in the tip.
Then the Nazis swept across Europe. Bíró sold everything he owned and fled with his family to South America.
Bíró set up shop in Argentina, where his pens became a hit.
BIRO PENS
An American named Milton Reynolds happened across them in Buenos Aires and brought a few back to the United States.
Imitation is the sincerest form of flattery.
Right?
Reynolds copied Bíró's design outright and mass-produced the pens.
Reynolds's pens became a phenomenon, helped by the huge crowd at Gimbels department store.
They also beat Bíró's pens to the American market by a matter of months, to the distress of Bíró.
But I escaped from Nazis....
It's fine! I'm flattering you.

New companies popped up, eager to get in on the trend.
SHEAFFER
EVERSHARP
Reynolds
PARKER
Marketing claims became increasingly ridiculous.
My pens don't need refilling for two years!
Mine write underwater!!
Mine write upside down!!!
Mine shoot LASERS!!!!
Wait... what?
Despite the hype, ballpoint pen technology at the time was really terrible.
The pens leaked all over the place and tore paper.
Ballpoint PENS
Over the next three years, the value of a ballpoint pen dropped from $12.50 to 50¢.
By 1951, ballpoint pens seemed doomed.
splort!
splut!
Hrr...
rrip!
Reynolds sold his business.
Welp, time to cash out.

Better ballpoint pens arrived eventually, and the public started buying them again.
New IMPROVED Ballpoint Pens
Better this
"Less tearing
Only this time, there weren't any crowds.
GIMBELS
PENS
What the heck were we all so excited about?
BRIEFER HISTORIES
In several countries, including Britain and Australia, ballpoint pens are known as Biros—after László!
Small victory, but I'll take it.
In 1965, a "space pen" which could work in zero gravity was developed by Paul Fisher. NASA bought 400 of them for $6 a pop.
Aggressive pen salesmen in the 1950s would scribble on clients' shirts then guarantee to buy their victim a new shirt if the ink couldn't be washed out!
A Bic ballpoint pen has enough ink in it to write a line two miles long. You'd probably misplace it first.

PAPER
Around the 1st century AD in China, a eunuch named Cai Lun invented modern paper using bark, hemp fiber, broken fishnets, and rags.
Hey, us eunuchs gotta be resourceful.
Cai was rewarded by the Emperor with a noble title and vast riches.
Oh, and, uh, sorry about your genitals.
But Cai eventually killed himself with poison when a bloody palace intrigue he was involved in went terribly awry.
Perhaps, I got a bit too resourceful....

The paper-making process was kept a closely guarded state secret of the imperial artisans.
I mean, think about what would happen if this fell into the wrong hands?
Useless comic-trivia books everywhere!
Finally, six centuries after Cai Lun offed himself, a battle in Central Asia let the cat out of the bag.
In 750 AD, revolution broke out in the Muslim Caliphate. A faction called the Abbasids seized power. The new rulers of the Islamic empire turned their attention eastward.
Hmm...there's this whole continent over there...
In 751, Arab armies slammed into the edges of the Tang dynasty, who ruled China.
Central Asia ain't big enough for the both of us.
BYZANTIUM
ABBASID CALIPHATE
UYGHUR KHAGANATE
TIBETAN EMPIRE
TANG DYNASTY
INDIAN KINGDOMS
NUBIA
Uh guys...
...guys?
The Caliph's army won the battle when two thirds of the Chinese Emperor's troops suddenly defected.
But the Abbasids got far more than just a victory....

Among the prisoners of war were two expert Chinese paper-makers.
Wait! We can show you a secret!
Uh... you have any broken fishnets?
A paper mill was set up in the city of Samarkand and the process spread through the Muslim world.
In Baghdad, the Abbasid capital, scholars bound sheets of paper together into books.
%$#&ing hipsters!
Large amounts of information were suddenly easier to share than ever before. The Islamic Golden Age flowered.
Mathematicians, philosophers, and scientists across the Muslim world exchanged theories and ideas.
Greek classics were translated.
And it actually became financially viable to be a full-time author or bookseller!
By the middle of the 9th century, the library of Baghdad held the largest number of books in the world and was called "The House of Wisdom."
Ooh! Killer name.

Then, in 1258, the Mongols swept out of the East, where they'd already been wreaking havoc for almost 50 years...

PENCILS
PENCIL
In the 19th century, the best pencils in the world were made from the wood of the American red cedar tree.
Ah, the great outdoors!
CHOP

The manufacturing process was wasteful. Only the knot-free heartwood, about a fifth of a typical tree, could be used to make pencils.
Maybe we should just set the rest of the wood on fire?
But not for warmth, only to laugh at.
Luckily, red cedar grew all over the American South.
The wood was so common, farmers used it to build barns and fences.
And we use more than just a fifth of the damn things.
Millions and millions of pencils were produced in the U.S. for export all around the globe.
CHOP! CHOP!
Good thing there's, like, infinite cedar trees.
CHOP! CHOP! CHOP! CHOP!
By the turn of the century, red-cedar supplies were dwindling.
Well, who could have *possibly* foreseen this?
People panicked.

Forests of red cedar were planted in Bavaria by German pencil barons but the wood warped in the alien climate.
Scheisse!
Red cedars were discovered on Little St. Simon's Island in Georgia. Pencil companies bought up the entire island, but found that the wood was subpar.
And now we own an island.
Great.
As wood supplies ran critically low, industry agents called "Cedar Cruisers" fanned out across the country.
Pen—who? You got three minutes to explain yerself or git off my property.
Ulp.
They bought up old fence posts and barns made from red cedar, then sent them back to be reprocessed into pencils.
Both pencil manufacturers and the U.S. Forest Service searched for alternatives.
Resource crises make for strange bedfellows.
Around 1925, a substitute was found! Incense cedar made good pencils, was abundant in the mountain ranges of the West and was easy to grow.
Great! Can we buy the whole mountain range?

There was only one problem.

To this day, the pencils we use are dyed to look like trees that they haven't been made from in almost a century.

BRIEFER HISTORIES

Britain banned pencil sharpeners during WWII because they were wasteful of wood and graphite.

Almost all graphite for pencils came from Barrowdale, England, until a new vein was discovered in Siberia and a reindeer-based mining operation was set up to extract it.

SNORT!

Thomas Edison loved shorter-than-regular pencils, which he had made special for him by a pencil factory.

The introduction of erasers panicked teachers who believed students wouldn't learn as much if they could correct their errors.

POST-IT NOTES
REMEMBER
TO THINK
OF A
JOKE FOR
THESE POST-ITS!
In 1968, Spencer Silver, a chemist at 3M, invented a glue that didn't stay stuck.
un-STICK!
Hm!
Silver thought there must be some way the glue could be useful. But for the life of him, he couldn't figure out how. He turned to his colleagues for help.
It doesn't stay stuck, Spence.
That's the POINT!
un-STICK
He didn't have much success.
Silver spent the next six years trying to figure out what problem the glue could be the solution for.
Listen, let me talk to you about this glue....

Arthur Fry, a product developer who also worked at 3M, had problems of his own.
Well... "problems" might be a bit of an overstatement.
But I'll concede that "minor choir-related annoyances" doesn't have gravitas.
Fry sang in a church choir and hated how the bits of paper he used to mark hymns fell out of his hymnal so easily.
flutter
In 1974, Fry attended a lecture given by Spencer Silver about his unsticky glue.
un-STICK!
DING!
You can just stick it, peel it off and stick it again!
yawn!
Silver's solution had found Fry's problem.

Fry coated one side of his bookmarks with Silver's glue. They stuck, but could be easily removed and reapplied.
I wonder if there's a use for this outside of choir-related activities.
un-STICK
Fry brought his idea to Silver.
It took the pair six years to develop the idea into a viable product.
Listen, let us talk to you about these bookmarks....
...a product that 3M had no idea how to market.
In 1977, the product was introduced as "Press 'n Peel." It was a huge flop.
It doesn't stay stuck.
SALE! 50% 75% off!
"Press 'n Peel" was pulled.
In 1980, the product was reintroduced as "Post-it Notes."
Idaho
Charge!
Boise
This time around, 3M had a plan — "The Boise Blitz."
The strategy involved giving away huge quantities of Post-it Notes in Boise, Idaho.
Hey, if we can make it here, we can make it anywhere.
FREE!

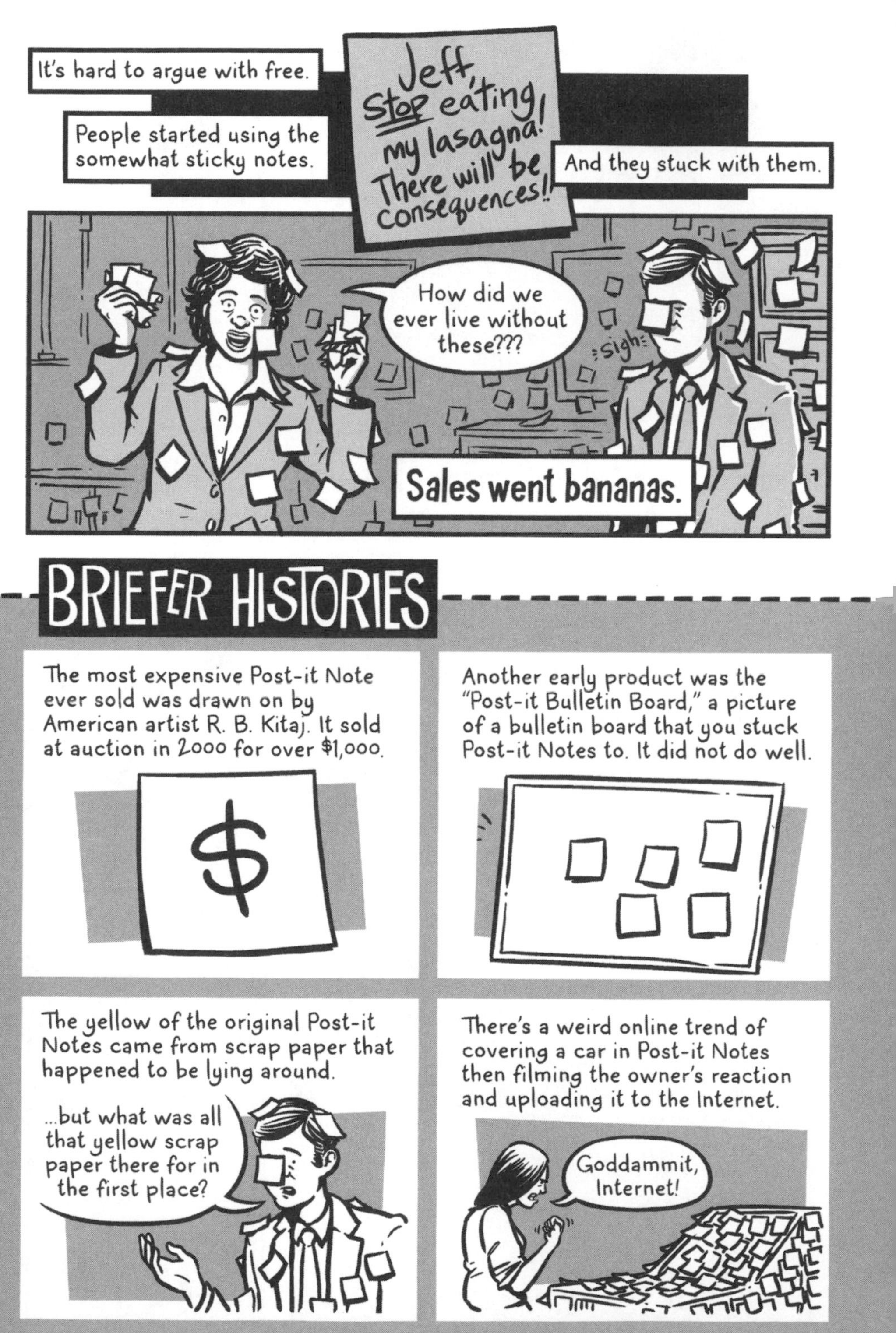

It's hard to argue with free.
People started using the somewhat sticky notes.
Jeff, STOP eating my lasagna! There will be consequences!!
And they stuck with them.
How did we ever live without these???
sigh
Sales went bananas.
BRIEFER HISTORIES
The most expensive Post-it Note ever sold was drawn on by American artist R. B. Kitaj. It sold at auction in 2000 for over $1,000.
$
Another early product was the "Post-it Bulletin Board," a picture of a bulletin board that you stuck Post-it Notes to. It did not do well.
The yellow of the original Post-it Notes came from scrap paper that happened to be lying around.
...but what was all that yellow scrap paper there for in the first place?
There's a weird online trend of covering a car in Post-it Notes then filming the owner's reaction and uploading it to the Internet.
Goddammit, Internet!

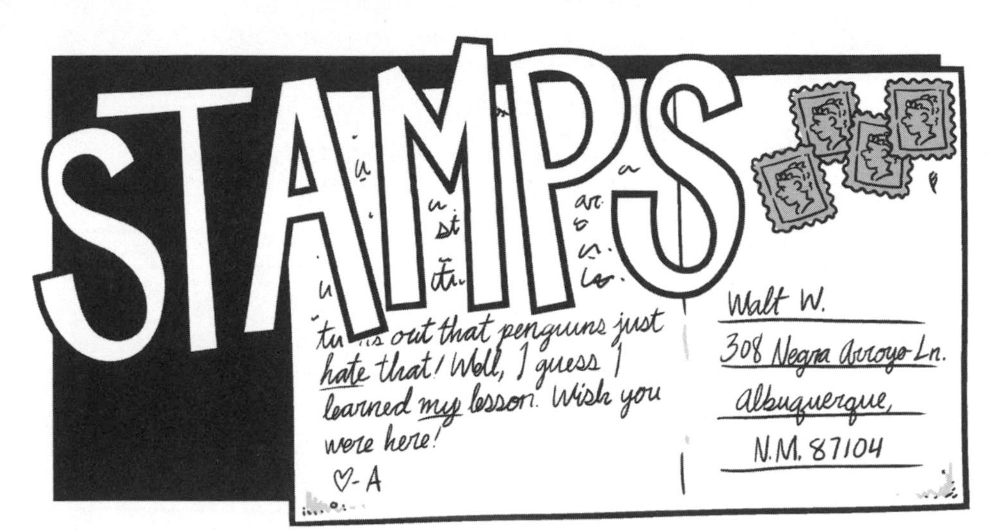

Abusing the postal system for a laugh was something that came quite naturally to Theodore Hook.

In 1810, Hook sent out thousands of letters requesting services and deliveries, all for the same day, and all in the name of Mrs. Tottenham of 54 Berners Street.

The resulting chaos caused enough of a traffic jam to bring a significant part of London to a screeching halt.
Tee-h... Uh. *cough* I should probably go.
CHAOS!
The prank became notorious as "the Berners Street Hoax." Hook was suspected, but never brought to trial.
Thirty years later, in 1840, the world's first lickable postage stamp was introduced in the United Kingdom.
It cost one penny and was black.
The stamp was dubbed "the Penny Black."
POSTAGE
M ONE PENNY. H
Suuuper imaginative name, you guys.
Queen Victoria!
Theodore Hook wasn't about to let a momentous event like the establishment of the modern postage stamp go off without a good prank.
Tee-hee! I should think not!

Front
Penny
OFFICIAL
Penates
Hook hand-painted a card with a caricature of postal workers huddled around an "official" inkwell.
He also wrote "Penny" and "Penates" (Latin for household gods).
Theodore Hook Esq.
Fullham
Back
It's not entirely clear why. He was possibly mocking the new stamp as something to worship.
He slapped a penny black stamp on the back, and addressed it to himself.
People love Latin jokes!
I am SO droll!
Post Of
And so the joke card made its way through the postal system...
Guess who....
...and returned to its sender...
Mail call... dick.
No—hee-hee —it's Latin— hee-hee!
You should see your face!

...and with that, Theodore Hook accidentally invented the picture postcard.
Hook's post card sold for almost $50,000 in 2002.
A bit of a price increase over a penny black!
POST CARD
POSTAGE
ONE PENNY
Tee-h... Uh, I did what?
BRIEFER HISTORIES
When he wasn't pranking the postal service, Theodore Hook was also a composer of comic plays and a respected satirist.
But messing with postmen was my true love!
Postcards with racy cartoons were popular in the 1930s. Sixteen million were sold a year in the United Kingdom until they were banned in the 1950s.
"If you were a doctor, I could show you something that would astonish you!"
Collecting postcards is known as deltiology, which is Greek for the study of small writing tablets.
Another stamp was introduced at the same time as the penny black, called the "two pence blue." It was blue, and cost two pence.
Again, killer name.
POSTAGE
TWO PENCE

THE GROCERY STORE
Paper Bags pg. 134
Instant Ramen pg. 138
Canned Fruit pg. 142
INSTANT NOODLE
INSTANT NOODLE

Ice Cream Cones pg. 146
Potato Chips pg. 150
POTATO CHIPS

PAPER BAGS
Wobble wobble
Paper bags in the 19th century were shaped like giant envelopes.
That meant that if you placed them on the floor, they would tip over and spill out their contents.
#!%@!
Well, this seems like a dumb way to do things.
flop flop flop
Margaret Knight was on the case.
Knight thought that flat-bottomed bags would be better, but needed a way to mass-produce them.
SCRIBBLE SCRIBBLE
M. Knight's Notes on Bags

Margaret Knight was no stranger to mechanical invention.
She'd grown up working in textile mills.
CHUCK CHUCK
Hooray. Childhood in the 19th century.
CHUCK CHUCK
While she was still a kid, she'd witnessed a fellow worker get maimed by a loose shuttle.
CHUCK THWIP!
Aiiiiee!
SPLORT
So, at the age of 12, she invented a stop-motion device to prevent it from happening again.
SCRIBBLE SCRIBBLE
By the time she was a teenager, it was adopted across the industry.
CHUCK
Hooray.
CHUCK CHUCK
But Knight never patented the device and saw no profit from it.
Still— less maiming.
But back to bag folding! Knight sketched out a plan and set off to a machine shop to have her contraption built.
Ye Olde Machine Shoppe

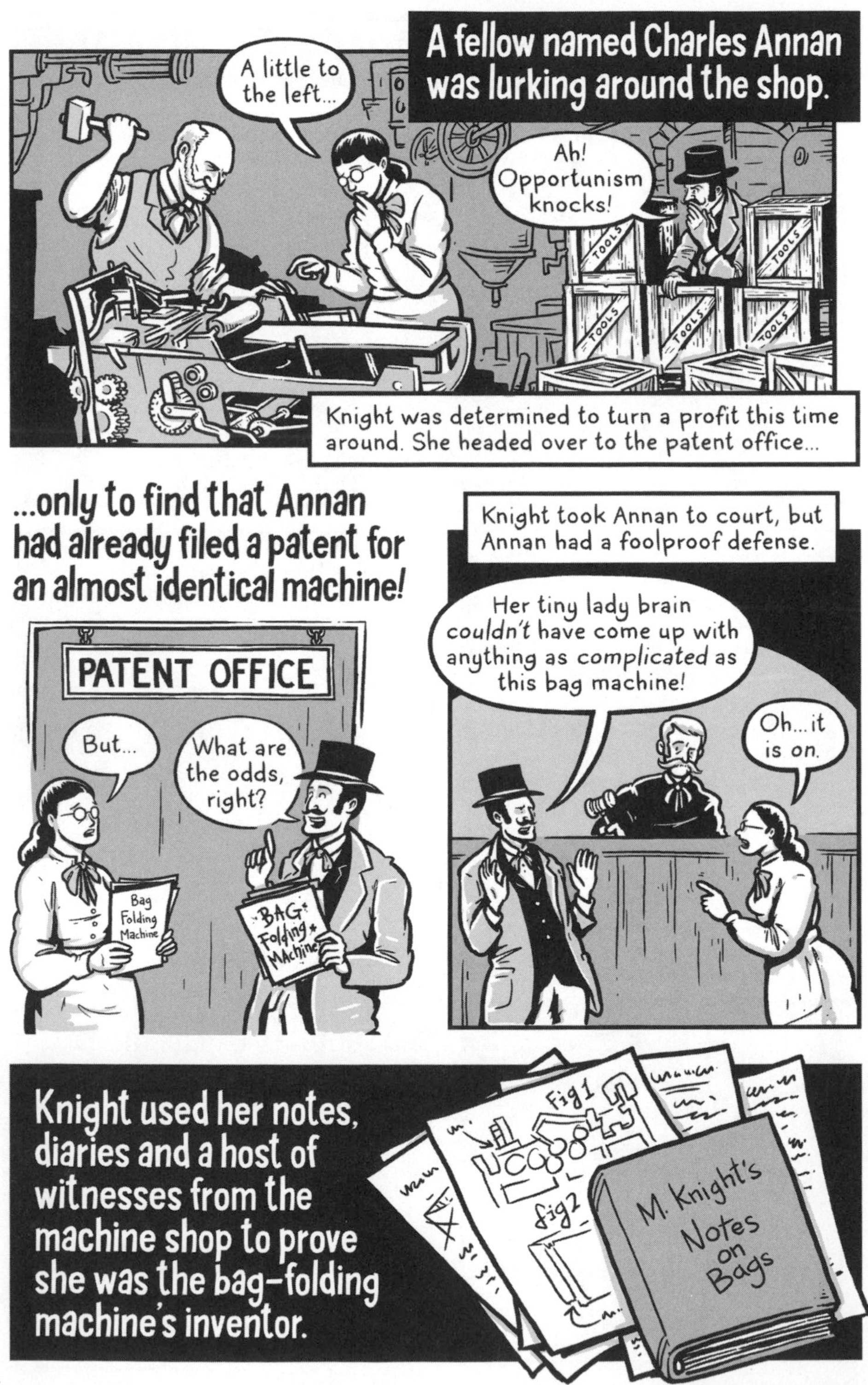
A fellow named Charles Annan was lurking around the shop.
A little to the left...
Ah! Opportunism knocks!
TOOLS
Knight was determined to turn a profit this time around. She headed over to the patent office...
...only to find that Annan had already filed a patent for an almost identical machine!
PATENT OFFICE
But...
What are the odds, right?
Bag Folding Machine
BAG Folding MAchine
Knight took Annan to court, but Annan had a foolproof defense.
Her tiny lady brain *couldn't* have come up with anything as *complicated* as this bag machine!
Oh... it is on.
Knight used her notes, diaries and a host of witnesses from the machine shop to prove she was the bag-folding machine's inventor.
fig1
fig2
M. Knight's Notes on Bags

In 1870, the court ruled in her favor.
BANG!
Tiny lady brain THAT, sucker!
I'll be going.
I have lurking to do.
cough
EVIDENCE
EVIDENCE
EVIDENCE
That made Knight the first American woman to win a patent suit and assured her place in history as the mother of the modern paper bag.
BRIEFER HISTORIES
Margaret Knight went on to receive 27 patents on everything from improved combustion engines to shoe-cutting machines before she died at the age of 76.
Just try me, Annan.
Just... try... me.
Knight's paper-bag-folding machine is on display at the Smithsonian National Museum of American History in Washington, D.C.
Paper bags are useful for ripening fruit because they trap and concentrate the ethylene gas that the fruit is releasing. That chemical changes color, texture and flavor.
Americans use up to 10 billion paper bags a year — which all adds up to about 14 million trees.
Ah, the great outdoors!
CHOP! CHOP!

INSTANT RAMEN
NOODLE CUP
It was 1956, and Momofuku Ando, a Taiwanese immigrant to Japan, was bankrupt.
Umm...could you put it on my tab?
slurp!
glorp
ラーメン
He'd tried selling prefab houses.
50% OFF!
Socks.
Engine parts.
He'd worked as manager for a credit association.
It went bankrupt.
A few lawsuits and a stint in jail for tax evasion had left him almost penniless.
%$#&@!
Maybe I should go back to engine parts...

Momofuku Ando needed a new business plan.
He remembered the food shortages from a decade earlier in postwar Japan.
Gurgle...
Ando resolved to dedicate his life to finding a way to feed everybody in the world.
Specifically, feed everybody with noodles.
glorp
slurp
He spent an entire year in his backyard garden shed experimenting with ways of prepackaging ramen.
Hey—great things happen in garden sheds!
His methods were a bit unorthodox.
He tried using a watering can to sprinkle his mushy noodles with chicken soup.
Result: even mushier.
In 1958, he developed a system.
You gotta have a system!
1 Ando boiled the noodles,
BLUB BLUB
2 then fried them,
SIZZLE!!
3 dried them,
DRY
PSSSS!
4 and then, deep-fried them.
This left tiny holes in the dried noodles, letting them reconsitute themselves quickly when hot water was applied.

Ando's noodles were initially marketed as a luxury item!
INSTANT COOK CHIKIN RAMEN
即席
チキンラーメン
特製
Prices soon dropped, though, and instant ramen noodles became popular in Japan.
I came to understand that all of my failure—all of my shame—was like muscle added to my body.*
*Actual quote!
Ando was a success.
And yet... my destiny feels strangely unfulfilled....
So in 1966, he turned his attention to feeding the entire world.
With noodles!!!
On a trip to the U.S., he'd seen people breaking instant noodles in half, sticking them in coffee mugs and eating them with forks instead of chopsticks.
Laziness in food preparation is the American way.
glorp
slurp
Inspired yet again, Ando began to package his instant noodles in Styrofoam cups.
He's sort of the patron saint of college students.

Cup Noodles became one of the most ubiquitous foods in the world.
glorp
slurp
Mankind is Noodlekind.*
slurp
glorp
Anywhere you were, as long as you had hot water and a fork, you had a meal.
*Ditto!
Ando died at 96. He claimed to have eaten chicken-flavored instant ramen almost every day of his life.
BRIEFER HISTORIES
Ando chose chicken as the original ramen flavor because it didn't violate anybody's religious taboos.
God loves chicken!
There's an Instant Ramen Museum in Osaka, Japan, where you can create your own cup of ramen using Ando's original ingredients.
The Instant Ramen Museum
CUP NOODLE
Ando's noodles traveled to space in 2005, two years before he died. They were a special zero-G variety that Ando called "Space Ram."
Space noodles are a pretty good swan song.
Ramen is universal, but flavors vary from country to country. Examples include bacon potato, taco, and something called "meat king."
I'm not sure I want to know what that is.
MEAT KING

CANNED FRUIT
Tasty
FRUIT COCKTAIL
IN HEAVY SYRUP
In 1872, the spirits of the dead revealed a method for canning uncooked fruit to Amanda Theodosia Jones.
Did she say uncooked fruit?
Jones had always had more than a bit of the mystic about her.
She claimed she could see the future, was visited by ghosts, and predicted her own father's death.
Turns out to not go over well at family dinner...
A spirit named Dr. Andrews told Jones to go to Chicago, where she got work as an editor.
When she fell sick, he came to her at night and cured her with magnetic currents.

The ghosts told Jones she was destined for great things.
Vague but great things.
But listen, I want to talk to you about that uncooked fruit canning....
Jones had visions of helping women, but no money to make that happen. Then came the canning tips from the great beyond....
With supernatural help, Jones devised a new canning system using vacuums and glass jars.
Fig. 1.
Fig. 2.
The spirits helped Jones out with legal documents and advice, but she claimed credit for the process, which she patented.
This $%#& is MINE!
Patent
Jones struggled to get her machines working, secure funding, and navigate the treacherous world of business.
And then the spirit said, "Back to the uncooked fruit...."
cough
For a few years, she gave up on the canning dream entirely and wrote a lot of poetry.
Hmmm...
What rhymes with vacuum?

Then, in 1890, Jones founded the Women's Canning and Preserving Company. Her destiny was at hand.

Jones's method was sound. The Women's Company was inundated with orders and expanded production.

A tidal wave of money flowed in. Jones smelled a rat.

BRIEFER HISTORIES

Jones also channeled a spirit named Judge Evelyn to write a manifesto calling for the adoption of a "Crusade Constitution" system of government.

In *A Psychic Autobiography*, Jones called the investors who seized control of her company "goblins." She got them arrested, but they were soon released.

ICE CREAM CONES
Saint Louis! 1904!!
The WORLD'S FAIR!!!
More than 19 million people visited the city over seven months.
The Crystal Palace can get bent.
And ice cream?
In a hot St. Louis summer, ice cream was selling very well, thank you very much!

So well, in fact, that a young ice cream vendor had run out of serving bowls for his customers.
CE CREAM
Ulp!
Next door, a Syrian immigrant named Ernest Hamwi was selling *zalabia,* a flattened and baked waffle-like treat.
munch munch
WAAAAAAAAAA
ICE
Hamwi noticed his neighbor's crisis and sprang into action. He rolled a *zalabia* into a cone, and suggested that they enter into a partnership.
See...you take one tasty food—
—and you put another tasty food *INSIDE* of it.
GASP!
drip!
The cone caught on!
Hamwi kickstarted America's love affair with portable ice cream.

Except the story was a little more complicated....
The Kabbaz brothers claimed that it was them, not Hamwi, who first put ice cream in edible cones at the World's Fair!
glare
glorp!
glare
gloop!
But so did Abe Doumar! Doumar even claimed to have an original cone-making machine to prove it.
But David Avayou disputed this and said that *he* invented the cone at the fair!! Then Charles and Frank Menches claimed exactly the same thing!!!
glare
And anyway, Italo Marchiony was already serving ice cream in pastry cups a year before the fair even happened. And a cookbook from 1825 described tiny waffle cones.
glare
There's even a painting of women enjoying ice cream cones in France that dates to 1807.
And so on and so forth.

BRIEFER HISTORIES

The world's largest ice cream cone, in Gloucester, England, was 13 feet tall and supported 2,204 pounds of ice cream. It was so big that its toppings had to be catapulted at it.

In 1984, President Ronald Reagan approved a resolution, declaring July National Ice Cream month.
It's morning in America.

Vanilla is the most popular ice cream flavor in the United States. Real vanilla mostly comes from Madagascar. That's also where lemurs are from.
...Lemurs are so great.

A typical dairy cow produces about 56 pounds of milk a day—a little more than enough to make two gallons of ice cream.

POTATO CHIPS

American legend, by contrast, has the chips invented in Saratoga Springs, New York, in the 1850s.

SARATOGA SPRINGS

NEW YORK STATE

George Crum, a famous restaurant owner and the son of an African American jockey and Native Huron woman supposedly served them to piss off a fussy eater.

But Crum never mentioned chips in the biography he commissioned. The story could be total bunk.
Oh, owning a nationally famous restaurant as a black man before slavery even ended isn't impressive enough?
GEO. CRUM
The legend was popularized by the snack food industry in the 1970s.
Regardless of where they came from, it was Laura Scudder, "Potato Chip Queen of the West," who made potato chips into the king of snacks.
Before 1929, potato chips were served out from large barrels in grocery stores into open wax bags.
SQUISH!
Ack!
Gross!
SQUISH!
Potato Chips
URP..
Why are we eating these?
But after a couple days sitting in the barrel, the chips were stale or mushy.
Laura Scudder grew up in Philadelphia, worked as a nurse, then lit out into the West in 1910, where she became the first woman to pass the bar exam in Ukiah, California.
But your... lady brain?

In 1926, Scudder started a food-packaging plant in Los Angeles.
Agh!
Gross!
Why are we making these?
Spuish
Food Packaging
She decided to tackle the stale potato chip problem.
Mayflower POTATO CHIPS
15¢
Scudder took a waxed bag and used an iron to seal the potato chips inside.
The potato chips stayed crisp! Scudder advertised them as "the Noisiest Chips in the World."
Music to my ears!
CRUNCH
But no matter how perfect her fix was, Scudder was still a female business owner in the 1920s. It was no cakewalk!
She couldn't get her delivery truck insured because insurance companies didn't believe a woman would pay her premium on time.
INSURANCE
Tiny lady brains couldn't possibly remember such things.
It's a known fact.
So Scudder found a female agent who'd write the policy.
CRUNCH
After Scudder's crunchy chips cornered the snack food market in California, she employed that same agent to write the policies for her entire fleet of vehicles.
Not being a @#$%head can be surprisingly lucrative!
CHIPS
CHIPS

America and the world fell in love with bagged chips.

BRIEFER HISTORIES

Scientists have found that amplifying the sound of the crunch in chips through headphones actually creates the impression that they're more fresh.

William Kitchiner also invented "wow-wow sauce." It involved pickled walnuts, butter, port, and, of course, mushroom ketchup.

THE BAR
Toothpicks
pg. 160
Beer Cans
pg. 156

Ice
pg. 164
Billiard Balls
pg. 168

BEER CANS
BEER
In the 1800s, cans were made of iron. Opening them required the use of a hammer and chisel.
Well, stand farther back next time!
Eventually, thinner metals like tin and aluminum began to be used.
Canned beer finally appeared for sale on January 24, 1935, in Richmond, Virginia.
CONTENTS 12 FL OZ
KRUEGER'S
CREAM
ALE
COOL BEFORE SERVING
Might we suggest a national holiday?
glub! glub!

A little tool called a "church key" was used to punch two holes in the top of the can and get the beer flowing.
I do think of it as a sacrament of sorts....
Psssh!
Ermal Fraze of Dayton, Ohio, forgot his church key while on a picnic in 1959.
This system seems to have some unforeseen limitations....
Fraze tried opening his beer with the bumper of his car.
There has *got to* be a better way to do this.
Fraze spent the next few months inventing a pull tab top for cans.
Eureka!
Psssh!
His pull tab simply popped off and was easily tossed away.
Fraze patented the pull tab in 1963.
Psssh!
Psssh!
Psssh!
Psssh!
Psssh!
Psssh!
By the end of the decade, they were on almost every beverage can.

But by the early 1970s, so many pull tabs had been used and discarded that a crisis was brewing.
toss
toss
toss
toss
toss
The pull tabs were small, hard to see, and slow to degrade.
Beachgoers stepped on them barefoot.
Tch! Gnarly!
Children and animals swallowed them, with predictably dire consequences.
EMERGENCY ENTRANCE
ENTER
This system seems to have some unforeseen limitations....
A movement to ban aluminum cans altogether began to pick up steam.
A "press tab" was developed, but it involved sticking your fingers into sharp openings.
What could—
hic!
—possibly go wrong?
It didn't catch on.
Finally in 1975, the sta-tab was invented by Dan Cudzik.
Why no "y?"
B cuz!
PLEASE RECYCLE

Almost immediately, sta-tabs became industry standard.
Thanks to the design, over half a billion tons of aluminum tops were recycled alongside their cans, rather than being discarded.
Beachgoers' feet were safe again...
Boy, good thing we got rid of all those pull tabs, huh?
...for the most part.
BRIEFER HISTORIES
Dan Cudzik's sta-tab was featured at an exhibit called Humble Masterpieces at the Museum of Modern Art in New York City.
The vast collection of the Beer Can Museum of East Taunton, Massachusetts, dates to when can-collecting was a fad in the 1970s.
It was a weird time.
An alternate evolutionary branch of the beer can family called the cone-top can lasted from 1935 to 1960, then faded away.
BEER
About 475 billion beverage cans are produced every year. That's around 66 cans for every single human on earth.

TOOTHPICKS
In the 1860s, Charles Forster's work for an import business brought him to Brazil.
Whoa. Humidity. Jeez.
He found the Southern Hemisphere quite a bit different than his native Massachusetts.
But what impressed Forster the most wasn't the mixed-up seasons, or the new birds, plants, and animals. It wasn't the unfamiliar languages and foods.
The lack of clam chowder does seem like an oversight, though.
PROD
No, it was the *teeth* of the Brazilians that astonished Forster.
They aren't utterly rotten and hideous!
SHINE
Weeeiird.

The Brazilians picked their teeth with tiny pieces of wood after meals. Forster was intrigued.
My god!
I think I can get rich with these things!
flick
PICK PICK
He brought a box of the wood slivers back to the States, set up a manufacturing plant and waited for toothpick orders to roll in.
Preventing hideous, rotten teeth — sort of a slam dunk of a business model.
CHOP CHOP SAW
CHOP WHITTLE CHOP
WHITTLE SAW
Orders did not roll in.
GENERAL STORE
Look, our customers are not exactly clamoring for boxes filled with tiny pieces of wood.
Forster wasn't deterred.
"Not exactly clamoring for boxes filled with tiny pieces of wood."
I'll show him....
He spread rumors that he'd closed his business due to lack of demand.
Then Forster hired an army of young men and women around Boston to ask stores for the "out-of-production" toothpicks.
!!!
?
?
?

When demand had built up, Forster "put himself back in business." Stores across Boston ordered toothpicks.
I can't explain it, but they really seem to have caught on with the kids....
Imagine that!
Forster then hired more youths to buy up all available toothpicks from the stores and return them to him.
Ahem...
I'm just gonna write this off as part of the marketing budget.
WINK
$$$
Having sold out their stocks, the stores reordered toothpicks. Forster sold them the same ones back again.
Order more of those boxes filled with tiny pieces of wood!
You know you can just call them toothpicks now, right?
!?
!?
!?
!?
!?
After Forster artificially inflated demand for half a year, toothpicks finally caught on in earnest without the assistance of clandestine market manipulation.
Well, sure, but what'll us youths do now?
Uh... We could try loitering?

Forster died a wealthy man.

The *American Stationer Magazine* declared that he'd done "more for the teeth of America than any other man under the sweep of her eagle's wings."

Forster left his toothpick empire to his daughter, Charlotte, who lived glamorously in Los Angeles, then created a sensation when she threw herself out of the third-story window of a sanitarium.

She survived.

Forster's manufacturing plant in Maine churned out toothpicks for 116 years. In 2013, cheaper imported toothpicks drove it into bankruptcy.

BRIEFER HISTORIES

Toothpicks existed long before Forster came across them. An early toothpick was discovered in the royal tomb of a Mesopotamian king who lived around 3,500 BC.

Right before Charles I had his head chopped off during the English revolution, he gifted his personal gold toothpick to a friend.

ICE
In the sweltering summer of 1844, the Wenham Lake Ice Company opened a branch in London.
WENHAM LAKE ICE COMPANY
Ice?
sweat
As in frozen water?
sweat
sweat
sweat
Sounds... kinda nice?
sweat
A massive fresh block of their product was placed in the window.
The following morning, a new block was installed.
sweat
sweat
sweat
sweat
And each day the process was repeated.
sweat
sweat
sweat
sweat
sweat
sweat
The store propped a newspaper behind the ice block so that passersby could see that it was clear enough to read through.
Give me all of it.
ALL. OF. IT.
SLAM!
sweat
sweat
sweat
sweat
sweat

Heat-addled London went collectively bonkers for the stuff.

At the age of 23, Tudor got into the ice business.
Ice?
As in frozen water?
In November of 1805, he packed a ship's hull full of Massachusetts lake ice and sailed it to Martinique.
What could possibly go wrong?
There were no buyers. Most of the ice melted and Tudor took a huge loss.
Ice was a hard sell. A shipment to England melted in port while customs officials argued over its classification.
CUSTOMS
Well, I still say it's a vegetable!
Lucky for Tudor, shipping ice was low cost, since it could be insulated using surplus sawdust from lumber mills.
And the ice grows back again every winter!
For free!!
SAW
SAW
Tudor kept at it. Within four years, he was turning huge profits and shipping ice all around the world.
For a few decades, ice was America's second biggest "crop" by weight.
Wenham Lake
Captain! The hull is fi—AIIIEEE!
Splash!
Tudor's Wenham Lake ice was tops.

BRIEFER HISTORIES

India was Tudor's biggest ice market. He built huge, ornate storage houses in Calcutta, Bombay, and Madras, but sold them after the business collapsed.

In the 1960s, the Madras ice house became a museum celebrating Swami Vivekananda, who brought yoga to the Western world.

14
BILLIARD BALLS
In 1867, the *New York Times* published an editorial warning that elephants faced imminent extinction.
The cause of this slaughter? The game of billiards, of course.
To have the proper "bounce," billiard balls could be made of only one kind of material — ivory from elephant tusks.
Look, I don't set the rules.
Billiards was hugely popular in the mid-1800s. And as the game spread, demand for ivory grew.
Good thing there's, like, infinite elephants.

Billiard balls were a tricky business. Only one in 50 tusks was actually good enough to manufacture a ball.
I just don't see proper bounce-potential quality in this one.
But one could only judge tusk quality after the elephant had been shot.
cough Pity, that.
Toss!
Sigh
The solution? Shoot more elephants.
KA-POW! KA-POW!
Lots more elephants.
I don't like the way this is heading...
By the mid-1860s, England was importing a million pounds of ivory annually. With each tusk averaging 60 lbs., that accounted for over 8,000 dead elephants a year.
You're starting to sound like one of those "peak elephant" nut jobs. Honestly!
Slowly, reality set in.
Say, if we run out of elephants, we can't make any more balls.
No balls, no billiards.
Gasp! Something must be done!

Michael Phelan was the world's most famous billiards player. He wrote bestselling books about the game and sold equipment.
No more billiard balls is somewhat of a problem for my business model.
In 1863, he offered a prize of $10,000 for the discovery of an ivory substitute.
Two brothers from Albany, New York, Isaiah and John Wesley Hyatt, read about the prize in their local newspaper.
Say, Isaiah, any interest in $10,000?
Splort!
Phelan Prize
The Hyatt brothers were printers, and they worked with a lot of chemicals. They spent five years experimenting.
Ok, now try not to breathe in for the next ... hour or so?
Finally, they refined a material made from cotton treated with acids into celluloid, a firm but bouncy plastic.
Ah, the sound of money!
BOUNCE!
rolllllll.....
But not the *proper* bounce. The Hyatt brothers didn't win Phelan's prize.
Go stick your "proper bounce."
We spent five years on this!

But celluloid was cheap and durable. It soon became popular as a material for everything *but* billiard balls.
Combs
Handles
Instruments
Dolls
Dominoes
Even teeth!
Celluloid film made the early movie industry possible.
The New York Times jokingly speculated that all the celluloid everywhere might catch fire or explode.
Ack!
My molar!
BOOM!
The first age of plastics had arrived!
Meanwhile, elephants continued to be massacred for billiard balls over the next half century.
It wasn't until 1907 that a plastic called "Bakelite" was invented that replaced the ivory in the balls.
BRIEFER HISTORIES
Of course, Bakelite didn't stop elephant poaching either. The African elephant population has declined from tens of millions in 1900 to fewer than 500,000 today.
sigh...
Michael Phelan used his celebrity to standardize the equipment for billiards, and got rich selling the model of tables, cues, and balls that he'd specified.
It pays to set the rules!

THE GREAT OUTDOORS
Traffic Lights pg. 174
Roller Skates pg. 178

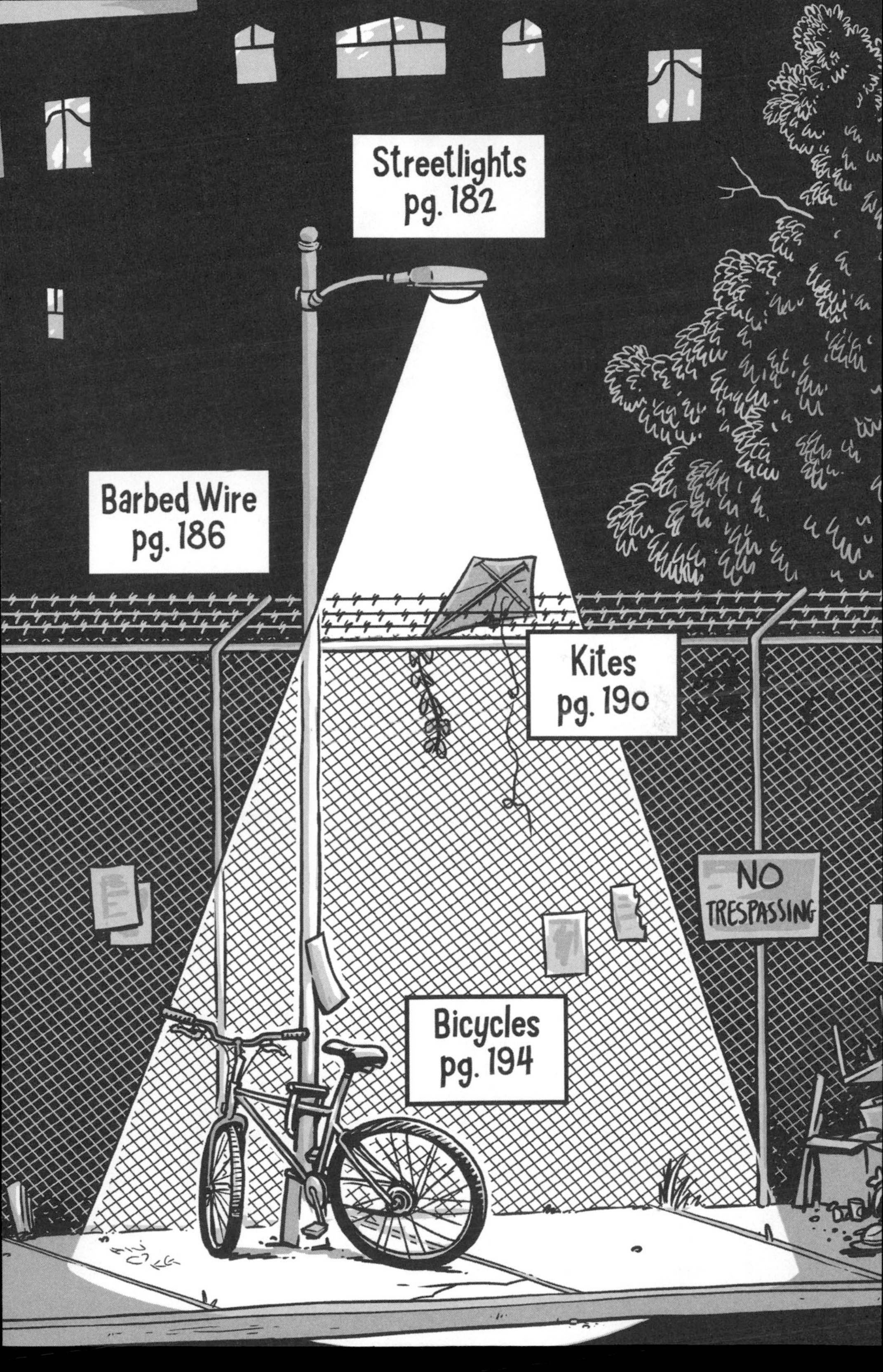

Streetlights
pg. 182
Barbed Wire
pg. 186
Kites
pg. 190
NO
TRESPASSING
Bicycles
pg. 194

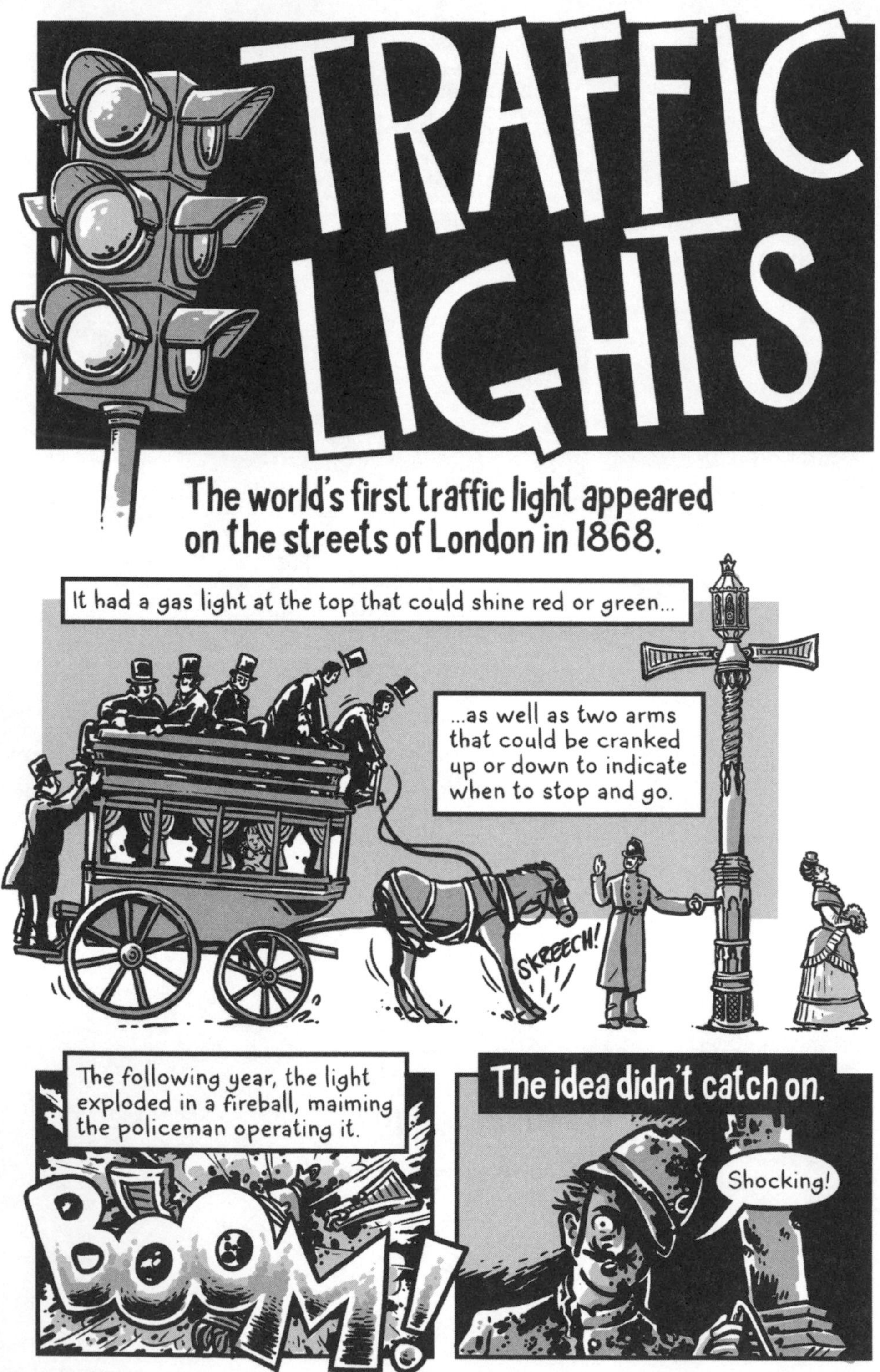
TRAFFIC LIGHTS
The world's first traffic light appeared on the streets of London in 1868.
It had a gas light at the top that could shine red or green...
...as well as two arms that could be cranked up or down to indicate when to stop and go.
SKREECH!
The following year, the light exploded in a fireball, maiming the policeman operating it.
BOOM!
The idea didn't catch on.
Shocking!

Garrett Morgan was a man who loved safety.
Safety and hair.
Morgan was born in 1877 to poor former slaves scraping by as sharecroppers in Kentucky.
I'm also a quarter Native American.
Reeeaally stacking the privilege deck.
He dropped out of school young and got work fixing sewing machines.
Morgan tried to come up with a liquid to polish sewing needles and accidentally invented a hair-straightening formula.
It's best not to ask why I poured the sewing-machine needle polisher on my head.
"IMPROVE YOUR APPEARANCE"
WAR DECLARED ON BAD HAIR
AGENTS WANTED TO CANVASS G. A. MORGAN'S HAIR PRODUCTS
G. A. MORGAN'S
COMB
$3.50
THE G. A. MORGAN HAIR REFINING COMPANY
Then he invented a new kind of comb and began selling hair products.
In 1912, at the age of 39, Morgan devised a hood to allow safe breathing while surrounded by gas or smoke.
This contraption became famous when Morgan used it to rescue miners trapped in a smoke-filled shaft deep under Lake Erie.
Wearing his safety hood, Morgan charged into the tunnel and dragged out the unconscious miners one by one.

The rescue became a media sensation and the safety hood caught on. Between that and the revenue from his hair products, Garrett Morgan became a wealthy man.
He invested some of that money in an automobile, becoming the first African American person in Cleveland to own one.
Safety, hair, and looking great in a convertible.
It had been a half century since the first failed traffic signal. With the advent of cars, roads were much more dangerous.
This piques one of my core interests!
SMASH
After witnessing a terrible accident at an intersection, Morgan decided to turn his attention to traffic signals.
In 1922, he developed a complex machine that used moving arms, warning bells and lights to control traffic.
DING! DING!
STOP
STOP
STOP
STOP
Ah, the elegance of simple design!
Morgan was a bit late to the game.
Two years earlier, a Detroit police officer invented a traffic signal that used three lights, similar to modern traffic signals.
ahem
glare
But the cop's signals hadn't yet spread to Cleveland. Technology traveled slowly then!

General Electric didn't care who'd pioneered what.
They bought Garrett Morgan's patent for an enormous sum of money...
...and traffic lights spread aross the nation.
Now richer than ever, Morgan devoted the rest of his life to political activism and publishing community newspapers.
Safety, hair, looking great in a convertible and good citizenship.
Everybody needs a hobby.
BRIEFER HISTORIES
Morgan founded the first black country club in Ohio, and then defended it from the Ku Klux Klan!
When marketing the safety hood, Morgan had a white friend pose as the inventor because racists didn't believe a black man could invent.
Even %#$&heads don't deserve to die of smoke inhalation.
Cleveland honored the memory of Morgan's rescue by naming a waterworks plant after him.
GARRETT A. MORGAN
WATER WORKS PLANT
The Detroit police officer didn't patent his traffic light and died in obscurity.
We've decided to move in together to save on rent.

ROLLER SKATES
A young Belgian clockmaker named John Joseph Merlin had a flair for the dramatic.
In 1773, Merlin built a silver clockwork swan automaton that preened and hunted fish.
CREAK!
Almost a century later, Mark Twain described it as possessing a "living grace."
PUFF
CREAK!
Merlin even tried to make a perpetual motion machine, but ended up with just a really weird clock that used atmospheric pressure to run.
tick tick tick tick tick tick tick tick
But before all that, in 1760, Merlin wanted to impress at a masquerade party.
I'm gonna look SO cool!
He attached wheels to the soles of his shoes.
He was 25 years old.

And probably without realizing it, Merlin invented the roller skate.
Oh, boy! Oh, boy! Oh, boy!
But it wasn't the innovation that excited the young man...
...it was the showmanship!
HEY EVERYBODY!!!
wobble
wobble
Merlin glided into the party on his skates while expertly playing the violin.
The guests were amazed.
gasp!
gasp!
gasp!

Unfortunately for Merlin, he'd neglected to also invent a brake for his skates.
WHAM!!
LOOK AT MEEEEEEEEEEEEEEEEEEEEEEE!!!
He gracefully sailed through the hall and crashed into a wall-sized mirror.
Merlin's novel footwear made an impression on his fellow partygoers, but perhaps not the one that he'd hoped for.
It's *ow* perfectly safe!
The roller skates didn't catch on.

...the era of roller skates had begun.

BRIEFER HISTORIES

In the late 1970s music, youth, poor choices, and roller skates reunited with the roller-disco fad.

Roller skates only appeared once in the Olympics, when they were used for roller hockey in 1992.

In 1956, Antonio Pirrello invented gas-powered roller skates that could roll up to 40 mph and were powered by a 19-pound backpack.

The largest collection of roller skates in the world is in the state of Nebraska, at the National Museum of Roller Skating.

STREET-LIGHTS
The first public electric lights were giant, super-strong bulbs stuck on top of huge towers scattered thinly across cities.
They were called "moonlight towers" and used arc lights, which were about 200 times as powerful as modern bulbs.
So... bright...
Many people hated them.
Health experts warned that exposure would cause eye diseases, nervous exhaustion, and freckles.
Gasp! FRECKLES!!

The bright, single-source lights left dark shadowed areas where they couldn't reach.
WHAM!
&%#$!
The moonlight-tower companies kept at it.
The future is coming, whether you like it or not!
ONLIGHT TOWERS INC.
They trumpeted the effect that arc lights had on crime reduction, circulating testimonials by burglars declaring that "electric lights are death to our trade."
Or at least they are where the light hits!
AIIIEEE!!!
Moonlight towers spread across America.
Detroit, eager to be the most modern city in the U.S., got on board in 1882.
Wisconsin
Michigan
Lake Huron
Canada
Lake Michigan
DETROIT!
Illinois
Lake Erie
City fathers installed seventy 150-foot-tall moonlight towers throughout the city.
We heard some really good burglar testimonials.
Not everybody was on board.
One man was arrested trying to chop down the tower next to his house.
Freckle-causin' devil moon pole!

The towers reportedly led to a mass die-off of sleep-deprived poultry due to the 24-hour illumination.
The future is coming, whether we like it or not.
Foggy nights thrust the entire city into darkness.
&%#$!
WHAM!
Towers collapsed in windstorms.
SNAP!
AIEEE!!!
Areas that were blocked by hills or tall buildings became dark and dangerous.
But...but your testimonials!
Detroit built more moonlight towers.
We're from the "just keep doing it again" school of problem-solving.
Then the city abruptly threw in the towel, dismantling all its moonlight towers after only a few decades of use.
You ever been in a bar when the lights come on? Imagine 30 years of that.
Across America, streetlights replaced moonlight towers.
The name's not as cool....
But fewer poultry die-offs is a pretty decent trade.

Despite their brief dominance over American nights, there are less than two dozen moonlight towers left today.

In 1970, the towers were designated Texas state landmarks.

BRIEFER HISTORIES

The first moonlight tower was erected in San Jose, California, in 1881. It was 237 ft. tall and was blown over by a storm in 1915.

Arc lights were used to light sets in the early days of film. The actors had to wear sunglasses between scenes.

BARBED WIRE
In the 1870s, barbed wire came to Texas.
What in tarnation?
Texas was a pretty crazy place back then.
"...back then?"
In the span of 30 years the region seceded from Mexico, became a U.S. state, was a battleground in the Mexican-American War, seceded with the Confederacy, then was reabsorbed back into the Union at the end of the Civil War.
UNITED STATES
Seems like a nice, quiet spot.
The Comanche, who controlled the Southwest, had been violently defeated.
New settlers brought cows.
LOTS of cows.
MEXICO

Cattle kings and cowboys grazed their herds on the open range.
Gettin' crowded.
When ranchers started using barbed wire to fence in those open ranges, the cowboys were not amused.
My grass.
MY! GRASS!
So they did something about it.
SNIP
In 1883, the "Fence-Cutting War" exploded.
Masked cowboy gangs with names like the Owls and the Blue Devils destroyed miles of barbed wire by night.
Why'd we call ourselves the Owls?
Owls are rad.
By the end of the year, more than 20 million dollars of damage had been done.
Ranchers hired armed guards.
His grass.
There were shootouts...
BANG!
BANG!
BANG!
BANG!
...and deaths.

But in September of 1884, the fence-cutters snipped the wrong barbed wire.

You didn't mess with Mabel Day.

The war flared up again briefly in 1886 when millions of dollars' worth of fences were cut in Brown County.

That did the trick, and the fence-cutting wars ended for good.

NO TRESPASSING

BRIEFER HISTORIES

Barbed wire was originally invented in 1868 by Joseph Glidden, who was inspired by a piece of wood with some spikes stuck in it.

There are barbed-wire museums all across America. One in Kansas displays over 2,000 different varieties of the stuff.

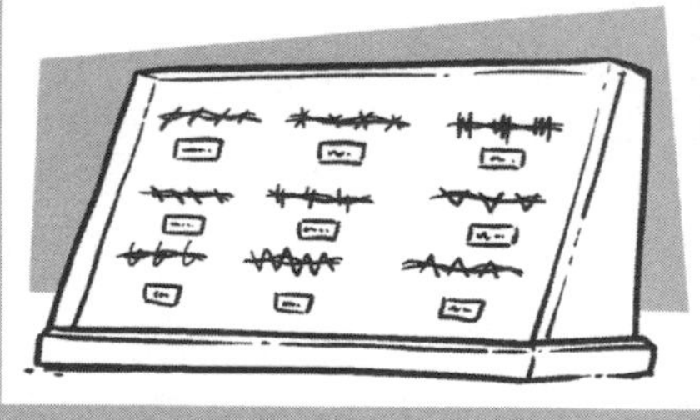

In the 1886 action, Texas Rangers went undercover as itinerant cowboys to find out where the cutter gangs would strike.

Barbed-wire saw widespread use as a brutal weapon of war in the trenches of WWI.

KITES
Nobody really knows for sure, but kites were probably invented around 2,500 years ago, and almost definitely in China.
Holy $%$#! Would you look at that!?
The Chinese had silk* and abundant bamboo. It was only a matter of time before someone put two and two together.
*See page 32!
This might have been the first time in human history that the power of flight was truly harnessed.
Hmm, now how can I use this to mess with my neighbors?
So the ancient Chinese used the technology to do what humans do best — kill each other.

In the Ming dynasty, crow-shaped kites filled with gunpowder were used to bomb enemy camps.
KA-BOOM!!
Crow bombs? Who said CROW BOMBS were allowed?
Kites were used to signal troops to attack.
Look, so pretty!
They were also flown to gauge how far men had to dig to get under castle walls.
Well, we seem to have gotten the measurements right at least.
Kites even helped establish China's mighty Han dynasty.
According to legend, when the dynasty's founder, Liu Bang, was trapped by a rival, he outfitted kites with noisemaking devices and flew them at night. Liu's enemies fled, terrified.
Huh!
That went better than expected.
KLAK! KLAK KLAK! KLAK KLAK! KLAK KLAK KLAK! KLAK KLAK! KLAK! KLAK! KLAK! KLAK! KLAK! KLAK! KLAK!

Kites were used for executions, too.
AIIIIIEEE
What? I'm inventing hang gliding!
After Emperor Wenxuan became the first ruler of the Northern Qi in 550 AD, he executed 700 of his political rivals by strapping them to giant kites and flinging them off the top of a very tall tower.
Emperor Wenxuan's interest in doing terrible things to people was far more broad than just kite-related cruelty.
Really, why limit yourself?
Gulp!
splat!
He also forced his soldiers to kill and eat one of their own lieutenants and publicly dismembered his concubine at a banquet.
If you're gonna do the whole "insane despot" thing, you gotta go all in.
Luckily, the emperor drank himself into the grave by the age of 33.
GURGLE
Oh, darn.
The emperor's dead.

Of course, humans didn't only use kites for war and torture. They also used them to steal from each other.

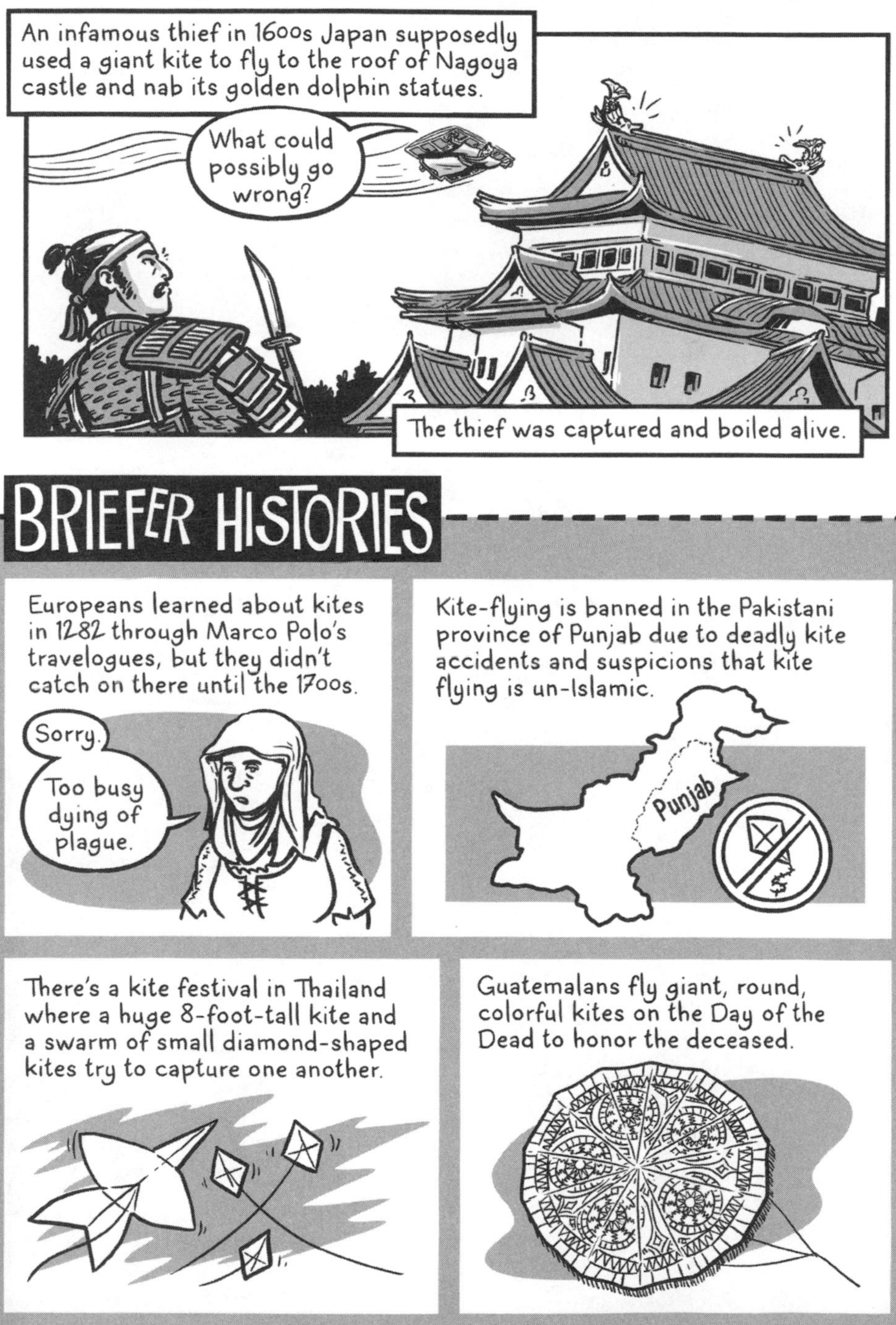

In 1896, Susan B. Anthony told a reporter that:

Millions of bikes were sold across the nation.
SOLD OUT!
FIRST NATIONAL
SHOW
THE NATIONAL BOARD OF TRADE
Huge exhibitions celebrated the cycle.
INDIANA BICYCLE
Promenading astride one was the height of fashion!
The bicycle craze took both sexes by storm, but the ladies soon realized that their constricting, Victorian clothes just weren't going to cut it.
Ow!
Gasp!
Christ!
Billowing petticoats got caught in wheels.
Ack!
Tight corsets restricted movement and breathing.
Wheeeze..
And hoop skirts?
&$%# hoop skirts.
For real.

American women responded with a whirlwind of invention.
Well, yeah, I mean, you read the last three panels, right?
SEW SEW SEW
Many bizarre garments resulted.
There were trousers with dress coverings you could clip on — a Victorian-era skort.
In 1892, Lena Sittig invented a sort of raincoat trouser suit called a "duck's back."
When you think about comfort and safety, think about ducks.
They sold faster than Sittig could make them.
I might... be on to something.
Quack!
Sittig followed this blockbuster with the "Duplex Bicycle Skirt," a huge overgarment that protected skirts from dust.
What can I say? I have a flair for product names.
The New York Times raved that it "transform[ed] the demure, darkly-gowned wheelwoman into the gayest of Summer maidens."
Too bad gay Summer maidens still get hella sweaty....
In all, about 30 women received patents on different inventions throughout the course of the craze.

Weird publicity stunts inevitably followed...

Londonderry Lithia Water Company paid Annie Kopchovsky $100 to change her name to "Annie Londonderry" and ride around the world on a bicycle.

Despite never having ridden a bike before in her life, she packed a change of clothes and a revolver and completed the journey in 15 months.

Meanwhile, doctors fretted that the bicycle's popularity was due to their seats leading young ladies to involuntarily pleasure themselves.

They didn't have long to fear. The bike mania died down by the turn of the century.

But women's fashions and fortunes had changed forever.

BRIEFER HISTORIES

The Susan B. Anthony bicycle quote was reported by another world-traveling woman named Nellie Bly. Bly journeyed solo around the globe in 72 days, set a world record and bought a pet monkey.

Bicycles were invented in 1817 by a German baron, and went through several totally insane-looking iterations before settling.

SHAM

BIBLIOGRAPHY

Bastone, Kelly. "The Sports Bra Turns 30." *Women's Adventure Magazine*, 2009.

Brady, M. Michael. "The Paper Clip Saga: The Invention That Was Not." *The Foreigner*, February 9, 2013. http://theforeigner.no/m/pages/columns/the-paper-clip-saga-the-invention-that-was-not/.

Braun, Sandra. *Incredible Women Inventors*. Toronto: Second Story, 2007.

Bundles, A'Lelia. *On her own ground: the life and times of Madam C.J. Walker*. New York: Scribner, 2001.

Burke, James. *The Pinball Effect: How Renaissance Water Gardens Made the Carburetor Possible—and Other Journeys*. Boston: Little, Brown, 1996.

Bryson, Bill. *At Home: A Short History of Private Life*. New York: Doubleday, 2010.

Deng, Yinke, and Pingxing Wang. *Ancient Chinese Inventions*. Cambridge, UK: Cambridge University Press, 2011.

Dien, Albert E. *State and Society in Early Medieval China*. Stanford, CA: Stanford University Press, 1991.

Dulken, Stephen Van. *Inventing the 20th Century: 100 Inventions That Shaped the World*. London: British Library, 2000.

Foner, Eric, and John A. Garraty. *The Reader's Companion to American History*. Boston: Houghton Mifflin Co., 1991.

Freeberg, Ernest. *The Age of Edison: Electric Light and the Invention of Modern America*. New York: Penguin Books, 2014.

Gates, Henry Louis, Jr. "Who Was the First Black Millionairess?" *The Root*, June 24, 2013. http://www.theroot.com/articles/history/2013/06/who_was_the_first_black_millionairess.html.

Gilbreth, Frank B., and Ernestine Gilbreth. Carey. *Cheaper by the Dozen*. New York: T.Y. Crowell, 1948.

Gillette, King C., *The Human Drift*. Boston: New Era Publishing, 1894.

Al-Hassani, Salim T. S., ed. *1001 Inventions: The Enduring Legacy of Muslim Civilization*. Washington, D.C.: National Geographic, 2012.

Hill, Louis. *Inventors and Inventions*. London: Black Dog, 2009.

"History of the Paper Clip." Early Office museum, n.d. http://www.officemuseum.com/paper_clips.htm.

James, Peter, and Nick Thorpe. *Ancient Inventions*. New York: Ballantine, 1994.

Jones, Amanda Theodosa. *A Psychic Autobiography*. New York: Greaves Publishing Co., 1910.

Kane, Joseph Nathan. *Necessity's Child: The Story of Walter Hunt, America's Forgotten Inventor*. Jefferson, NC: McFarland, 1997.

Kealing, Bob. *Tupperware, Unsealed: Brownie Wise, Earl Tupper, and the Home Party Pioneers*. Gainesville: University of Florida Press, 2008.

Keen, Catherine. "Jogbra: Providing Essential Support for Title Nine and Women Athletes." *O Say Can You See?* National Museum of American History, December 11, 2014. http://americanhistory.si.edu/blog/jogbra-providing-essential-support-title-nine-and-women-athletes.

Krell, Alan. *The Devil's Rope: A Cultural History of Barbed Wire*. London: Reaktion, 2002.

Levy, Joel. *Really Useful: The Origins of Everyday Things*. Buffalo, NY: Firefly, 2002.

Liu, Joanne S. *Barbed Wire: The Fence That Changed the West*. Missoula, MT: Mountain Press Publishing Co., 2009.

Macdonald, Anne L. *Feminine Ingenuity: Women and Invention in America*. New York: Ballantine, 1992.

Maxwell, D. B. S. "Beer Cans: A Guide for the Archaeologist." *Historical Archaeology*, 1993: vol. 27, no. 1., pp. 95-113.

Meyers, James. *Eggplants, Elevators, Etc.: An Uncommon History of Common Things*. New York: Hart Pub., 1978.

"A New Bicycle Skirt." *The New York Times*. October 15, 1893.

Petroski, Henry. *The Evolution of Useful Things*. New York: Knopf, 1992.

_____. *Invention by Design: How Engineers Get from Thought to Thing*. Cambridge, MA: Harvard UP, 1996.

_____. *The Pencil: A History of Design and Circumstance*. New York: Knopf, 1990.

_____. *Small Things Considered: Why There Is No Perfect Design*. New York: Knopf, 2003.

_____. *The Toothpick: Technology and Culture*. New York: Knopf, 2007.

Pilon, Mary. *The Monopolists: Obsession, Fury, and the Scandal behind the World's Favorite Board Game*. New York: Bloomsbury USA, 2014.

Seaburg, Carl. *The Ice King: Frederic Tudor and His Circle*. Massachusetts Historical Society, Boston, and Mystic Seaport, Mystic, Connecticut, 2003.

Smith, Robert A. *Merry Wheels and Spokes of Steel: A Social History of the Bicycle*. San Bernardino, CA: Borgo, 1995.

Sullivan, Otha Richard, and James Haskins. *African American Inventors*. New York: Wiley, 1998.

Temple, R., *The Genius of China*. New York: Simon & Schuster, 1986.

Vare, Ethlie Ann and Greg Ptacek. *Mothers of Invention: From the Bra to the Bomb: Forgotten Women & Their Unforgettable Ideas*. New York: Morrow, 1988.

——. *Patently Female: From AZT to TV Dinners: Stories of Women Inventors and Their Breakthrough Ideas.* New York: Wiley, 2002.

Ward, James. *The Perfection of the Paper Clip: Curious Tales of Invention, Accidental Genius, and Stationery Obsession.* New York: Touchstone, 2015.

Weightman, Gavin. *The Frozen-Water Trade: A True Story.* New York: Hyperion, 2003.

Weissmann, Dan. "How Billiards Created the Modern World." *Marketplace.* March 4, 2015. http://www.marketplace.org/2015/04/03/business/how-billiards-created-modern-world.

Willard, Frances Elizabeth. *A Wheel Within a Wheel—How I Learned to Ride the Bicycle: With Some Reflections by the Way.* New York: F.H. Revell, 1895.

Wilson, Bee. *Consider the Fork: A History of How We Cook and Eat.* New York: Basic, 2012.

Tasty
FRUIT COCKTAIL
IN HEAVY SYRUP
NET WT 1LB 14OZ

ACKNOWLEDGMENTS

My deepest thanks to Dakota, dw, Kathy, and, especially, Ben for reading the drafts of this book and offering their thoughts and edits. Thanks to my agent, Farley Chase, for helping me navigate the process of putting together a book. I owe the most to my editor, Anna deVries, who found me in the first place.

I'm grateful to my parents for packing my childhood home with trivia books I devoured and to my siblings for not rolling their eyes too much when I'd spend dinner reciting what I'd learned. But I couldn't have done any of this without my wife, Kathy, the best companion I've ever known.

BEER

NOODLE
CUP

ANDY WARNER's comics have been published by *Slate*, *The Nib*, *Fusion*, American Public Media, KQED, *Symbolia*, the United Nations Relief and Works Agency, UNICEF, and *BuzzFeed*.

He has taught cartooning at Stanford, the California College of the Arts, and the Animation Workshop in Denmark.

He writes and draws in a garden shed in San Francisco and comes from the sea.

andywarnercomics.com

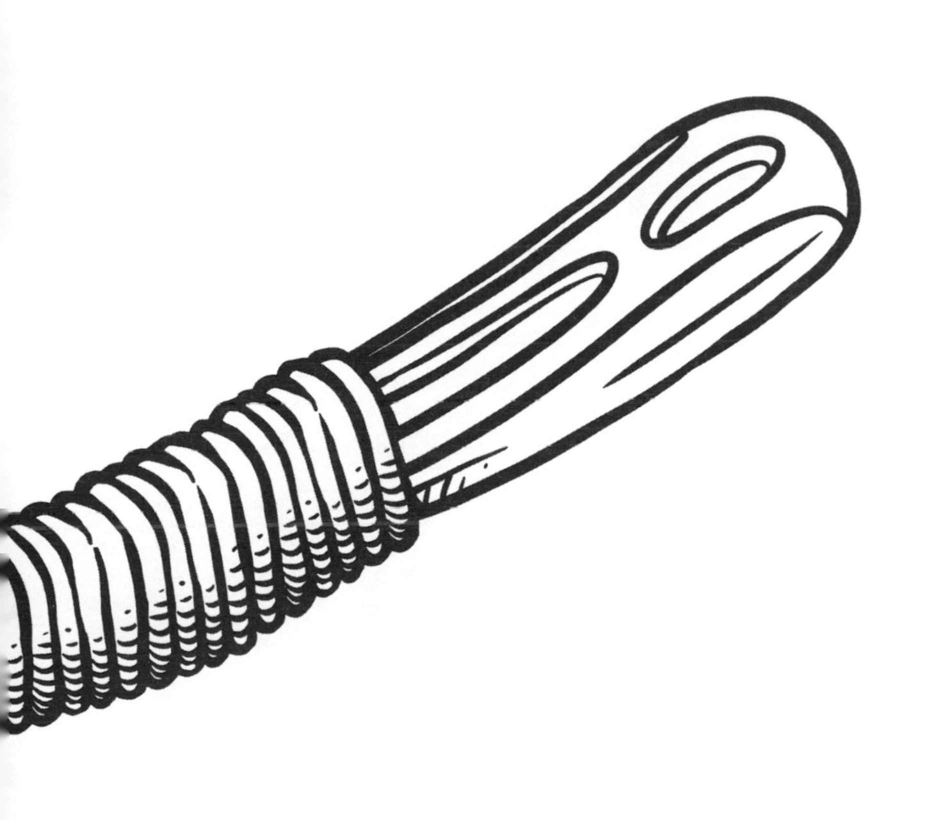

Tasty
FRUIT COCKTAIL
IN HEAVY SYRUP
14